Plants+People

An exhibition of items
from the Economic Botany Collections
in Museum No.1

Centre for Economic Botany
Education Department

The Royal Botanic Gardens · Kew

Introduction to the Economic Botany Collections

. . . plant products that are 'either eminently curious or in any way serviceable to Mankind'.

Kew's Museum of Economic Botany was founded by Sir William Hooker, when he became the first Director in April 1841. He was previously Professor of Botany at Glasgow Botanic Garden and brought with him the collections of specimens of textiles, gums, dyes and timber that he used to illustrate his lectures.

Hooker had obtained his post in Glasgow with the support of Sir Joseph Banks, Director of George III's botanic garden. Banks, like Hooker, had a particular interest in the economic exploitation of plants and it was he who commissioned Captain Bligh of the infamous HMS *Bounty* to transport seedlings of the breadfruit from Tahiti to the West Indies. There were two Kew botanists on the expedition but the voyage ended in mutiny, the sailors complaining among other things that the plants were better treated than they were. The breadfruit was successfully transported on Bligh's second voyage.

When Hooker first came to Kew, there was nowhere to house his collections until 1846, when the Royal Family gave up their kitchen and fruit garden and made available a building once used as a fruit store and as accommodation for the Royal servants (now the School of Horticulture). Here Hooker displayed his collections on trestle tables and successfully persuaded the Commissioners of Woods and Forests, who managed Kew at that time, of the value of a museum of economic botany. The architect Decimus Burton was asked to adapt part of the building for the collections, and the first Museum of Economic Botany in the world was opened on 20 September 1847. It was an immediate success.

The collections quickly grew with contributions from all over the world, due in large measure to the policy of the then Secretary of State for Foreign Affairs, who encouraged British consuls to submit specimens. The aim of this work was to further the economic development of the British Empire through, for example, the collection of cinchona (the source of quinine) and rubber seeds in South America which were then introduced into Southeast Asia.

Contributions from the Great Exhibition of 1851 in Hyde Park and the Paris Exhibition of 1855 added to the growth of the collections. An interesting and extensive group of items connected with the tea trade with India and China were also acquired at this time.

The design of Decimus Burton for Museum No. 1. Numerous windows on each elevation allowed all the showcases to be illuminated with natural light. Burton also worked on the design for the two major Victorian glasshouses and the Main Gate at Kew.

Many famous expeditions and surveys were taking place and Hooker was able to arrange for botanists to accompany some of these. Thus more specimens and artefacts reached Kew. Besides being Director of Kew, Hooker was an adviser to Queen Victoria's government on the cultivation and distribution of economic crops. He also influenced the setting up of a network of botanic gardens throughout the British Empire. Many well-known botanists and naturalists donated items, giving details of their local use, as well as information on their botanical and geographic origin.

William Hooker's son, Joseph, was in India from 1848 to 1850 collecting specimens in East Bengal and Sikkim. He sent many items to the Museum collections, notably objects related to the manufacture of opium. He sent teapots and brick tea from eastern Nepal, together with instructions received from Tibetans in its preparation. He eventually succeeded his father as Director of Kew and in all donated over 700 objects to the Economic Botany Collections.

By 1855 the size of the collections had outgrown the original premises. Visitor numbers were so great that William Hooker often found it impossible to gain entrance with distinguished guests. Decimus Burton was again asked to design a suitable building and the site chosen was across the lake, opposite the Palm House. This building was opened in 1857 and created roughly twice as much space as the first Museum (curiously called Museum No. 2, the second Museum being called Museum No. 1!).

The collections continued to grow. In 1878 the Government of India donated an important collection of forest produce. In 1880 the economic and botanic collections from the India Museum in South Kensington were transferred to Kew, necessitating the construction of an extension to the 1857 building.

William Hooker.

Joseph Hooker (seated far left) at an encampment in La Vesa Pass, Colorado, in the Rockies, in July 1877. He was accompanying a party of the US Geological and Geographical Survey and collecting plants with Asa Gray (on ground), Professor of Botany at Harvard University.

Another extremely important collector was Richard Spruce, a botanist who worked in South America from 1849 to 1864. He sent Hooker a remarkable collection of items made from plants by tribes living in Brazil, Peru and Ecuador, all accompanied by meticulous descriptions of their use. David Livingstone, one of Britain's most famous nineteenth century explorers, donated objects from Africa. Archaeologists working in Egypt sent some of the plant-based artefacts from tombs back to Kew for identification. Sir James Brooke was a soldier with the British East India Company. He became Rajah of Sarawak and collected both plants and plant products which he sent to Kew.

When trade with Japan was opened up. the British government showed a particular interest in the Japanese lacquerware and paper-making industries. Sir Harry Parkes was British Minister in Tokyo and in 1869 the then Prime Minister, William Gladstone, asked for a report on paper-making in Japan. In 1871, some 400 specimens of paper and paper products were sent from Japan to England and divided between Kew and the Victoria & Albert Museum.

In 1882 John James Quin, the Acting Consul at Hakodate, published a detailed report on Japanese lacquer-ware and sent an important set of tools used in tapping and the preparation and application of lacquer.

The success of the Museum encouraged various firms to donate specimens, in particular the rubber manufacturer Charles Mackintosh & Co. More recently, the Royal Pharmaceutical Society of Great Britain gave its important collection of natural medical materials. From 1841 it had acquired a collection of crude drugs. By the time the collection was donated to Kew in 1983, there were over 9,000 specimens.

The study of plants used by people, loosely called economic botany, has formed an important part of the work of the Royal Botanic Gardens, Kew since the time of George III. In the nineteenth century its aim was to aid the economic growth of the British Empire. Its focus today is on the sustainable use of plants and on conservation. The work is carried out in collaboration with colleagues in the countries of origin of the species being studied. Much important work is being undertaken in the search for potential sources of

KEY TO THE EXHIBITION

1 Central cabinet
2 Growing collections
3 Fabulous fabrics
4 Getting in a lather
5 Healing plants
6 Highs and lows
7 Pick-me-up plants
8 Sugar and spice
9 Taking our pulses
10 Eating for energy
11 Baubles, bangles and beads
12 Nature's bounty – coconut
13 Nature's bounty – pineapple
14 Head to toe
15 Hunting, shooting and fishing
16 To have and to hold
17 Words and pictures
18 Shake, rattle and blow
19 Playing with plants
20 Plants for energy
21 Plants on the move
22 To dye for
23 Where to go

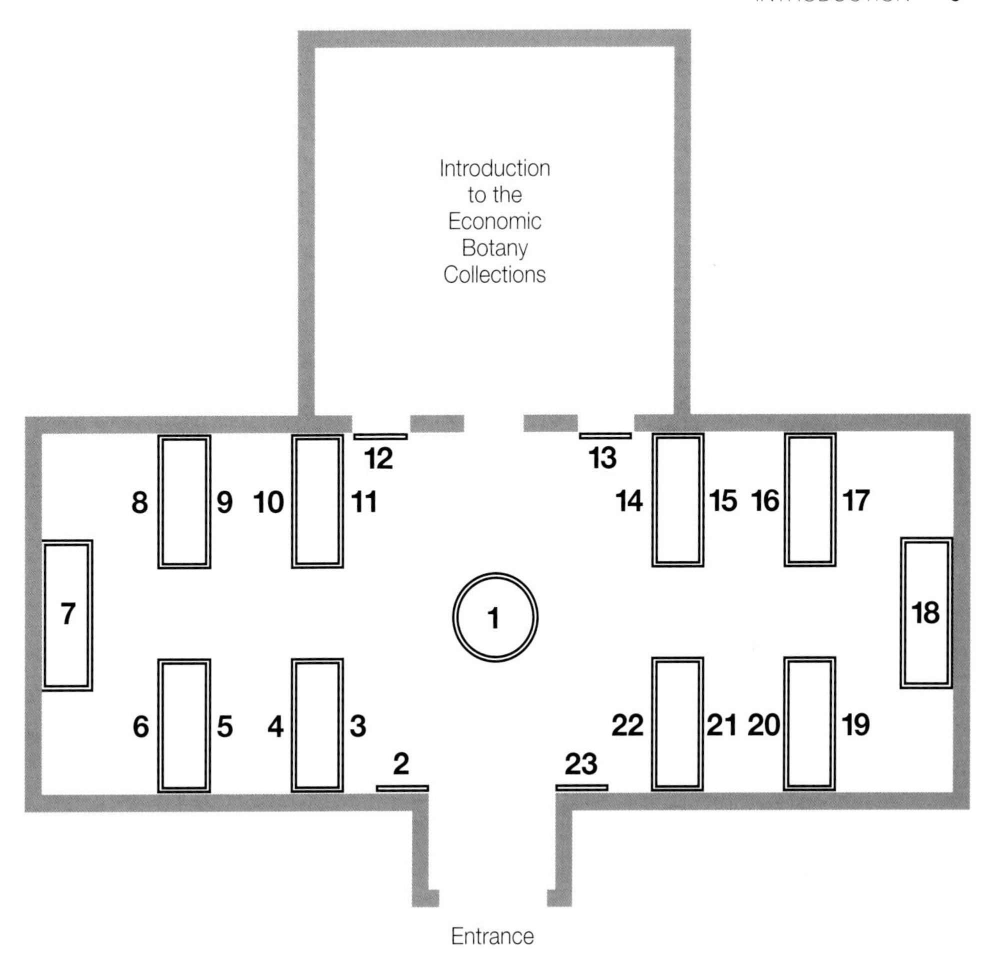

medicinal drugs. Another field aims to ensure that millions of people in Africa still have wood for their fires in the twenty-first century. Many of their staple foods are either toxic or unpalatable unless cooked. Without a sustainable source of firewood, people will resort to less nutritious roots and berries, with resulting potential problems of malnutrition. Work is being carried out to identify fast-growing native trees which are easy to harvest, can survive long periods of drought and can also perhaps provide extra benefits in the form of edible fruit, or feed for animals, medicine or even good timber for fencing or building purposes. The use of particular grasses as an important renewable source of energy is a further area of study.

The aim of much contemporary economic botany research is to find ways to manage the earth in a sustainable way, to provide for current needs and at the same time to conserve resources for future generations. Solving problems in economic botany requires a detailed knowledge of the evolutionary relationships between plants, their chemistry, natural habitats and methods of cultivation. The Royal Botanic Gardens, Kew makes an important international contribution to this work.

The Economic Botany Collections at Kew are the oldest and the most comprehensive of their kind in the world. They consist now of over 76,000 specimens and are continually being added to. The Plants + People exhibition gives visitors a glimpse of these Collections and emphasises Kew's international importance in this field. They serve as a timely reminder of the close relationship between plants and people, in all aspects of their lives. Kew is in a unique position to make a major impact on the future conservation, management and sustainable development of the world's plant resources.

In 1853 Hooker remarked that 'The Economic Botany Museum has done more to popularise . . . knowledge of the vegetable creation than all the palms, the gorgeous water-lilies, the elegant ferns, etc. . . . which grace the tropical houses of these noble gardens.' Today, it is the most outstanding and comprehensive museum of its kind in the world.

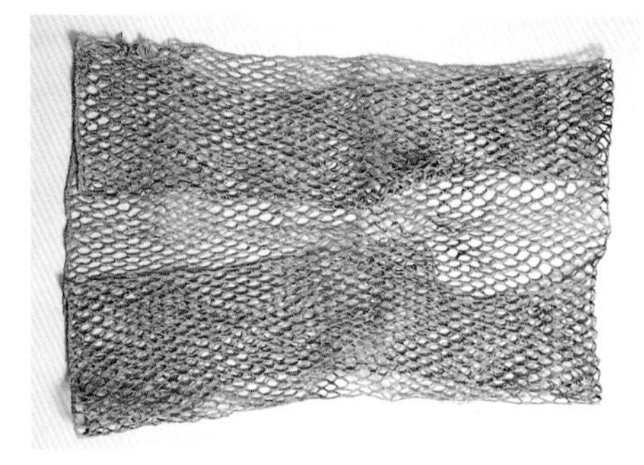
1

Some exhibits from Kew's first directors and others

Sir William Hooker, the founder of the Museum of Economic Botany, believed that it complemented the living plants in the Gardens. It was the first museum of its kind in the world.

Walking stick
Sugar cane *Saccharum officinarum*
Donated by J Banks
EBC 38757
Sir Joseph Banks (1743–1820), Kew's first unofficial director, had a passionate interest in useful plants. His name is commemorated in the building which today houses most of Kew's Economic Botany Collections.

Seed heads
Banksia dentata
Donated by G Pattison 1976
EBC 44991 Papua New Guinea
Joseph Banks accompanied Captain Cook on his voyage around the world (1768–71). He was particularly enthusiastic about the plants he encountered in Australia – one genus of Australasian plants, the banksias, was named in his honour.

Woven bag
New Zealand flax *Phormium tenax*
Donated by W J Hooker
EBC 30000 New Zealand

Ink
White maple *Acer* sp.
Donated by A Mountain
EBC 62232 Canada
Sir William Hooker received this ink from the Reverend Armine Mountain in Canada for his collection of useful plant products.

Wood
Bahama pine *Pinus caribaea*
Donated by Lieutenant Governor Nesbitt 1854
EBC 18667 Bahamas
Officials on service abroad sent samples to Sir William Hooker for his collection.

Brush
Garden cress *Lepidium sativum*
Donated by J D Hooker 1869
EBC 67344 Russia
Joseph Hooker (1817–1911), who succeeded his father, William, as Director of Kew, donated over 700 objects to the Economic Botany Collections.

Lepcha snuff box
Bottle gourd *Lagenaria siceraria*
Donated by J D Hooker
EBC 54628 Tibet
Joseph Hooker collected many Himalayan plants for Kew on his travels through Sikkim, Tibet and Nepal between 1848 and 1851. He also sent back artefacts for his father's collection of economic plants.

Sleeve studs
Arctostaphylos glauca
Donated by J D Hooker
EBC 51207 USA
During the twenty years he spent as Director of Kew, Joseph Hooker remained an enthusiastic plant collector. In 1877, he visited the Rocky Mountains of Colorado and Utah in the United States to study their flora.

Fish net
Securidaca longipedunculata
Donated by J Kirk 1859
EBC 66905 Zambezi, Central Africa
John Kirk was a member of Dr Livingstone's second Central African expedition (1858–63) following the course of the Zambezi river. This fish net was one of several objects that Kew received from him.

Cotton cloth
Cotton *Gossypium* sp.
Donated by J Brooke 1852
EBC 65620 Borneo
Sir James Brooke was a soldier with the British East India Company who became Rajah of Sarawak. As he explored Sarawak, he collected plants and plant products, which he sent back to Kew.

Baobab bark
Baobab *Adansonia digitata*
Donated by T Baines 1871
EBC 65249 South Africa
Thomas Baines suggested that this bark was as good a cure for fever as quinine. He travelled widely in Africa and Australia, painting accurate pictures of the plants and people he encountered.

Portion of a wreath
Blue waterlily *Nymphaea caerulea*
Donated by Dr Schweinfurth 1883
EBC 40729 Egypt
This wreath, from the coffin of Rameses II (c. 1200–1100 BC), consisted of parts of blue waterlily flowers together with leaves of *Mimusops schimperi* arranged on strips of date palm leaf (*Phoenix dactylifera*).

An early 1960s view of the first home of the Economic Botany Collections, the so-called Museum No. 2. It is now Kew's School of Horticulture.

Network coat made of paper (1)
Paper mulberry *Broussonetia kazinoki*
Donated by H Parkes 1871
EBC 42867 Japan
This coat, made of recycled paper, still shows the ink from a previous use. It was collected on behalf of Sir Harry Parkes, the British Minister in Tokyo, to illustrate his report on paper-making in Japan in 1871.

Model of temple
Vegetable ivory
Phytelephas macrocarpa
Donated by R Taylor 1865
EBC 36048
This object was exhibited by Benjamin Taylor (manufacturer) at the Great Exhibition of the Works of Industry of All Nations in 1851. Such intricately carved temples cost £7. Vegetable ivory seeds and objects made from them were illustrated in the *Official Descriptive and Illustrated Catalogue* of the Exhibition.

Model of linseed oil press
Donated by India Museum 1880
EBC 73046 India
In 1880, the Economic Botany Collections received many specimens from the India Museum, which had been created by the British East India Company to display artefacts their employees had collected.

Fine ornamental bottle of rubber
Rubber *Hevea brasiliensis*
Donated by C Mackintosh and Co. 1853
EBC 44181 Brazil
This bottle was donated by the manufacturing company of Charles Mackintosh, who had been one of the first people to make waterproof cloth by coating fabric with a thin layer of rubber.

Wooden elephants
1 *Albizia odoratissima* EBC 37879
2 *Mimusops hexandra* EBC 37900
3 *Eugenia aquea* EBC 37928
4 *Tectona grandis* EBC 37903
5 *Adina cordifolia* EBC 37931
6 *Tamarindus indica* EBC 37913
7 *Cassia auriculata* EBC 37898
8 *Pericopsis mooniana* EBC 37906
9 *Memecylon rostratum* EBC 37917
10 *Eugenia sylvestris* EBC 37921
Donated by F J Dawes
Sri Lanka
Among the many artefacts displayed in the Wood Museum (now the Kew Gardens Gallery) were these elephants made of Sri Lankan woods. They were collected by Frederick Dawes, an accountant who worked in Sri Lanka, and given to Kew after his death in 1966.

Cabinet of essential oils (2)
1 Essential oil from sage
Salvia officinalis
2 Sandalwood oil
Santalum album
3 Convallarine
Convallaria majalis
4 Barbaloine
Aloe sp.
5 Cephaline hydrochloride
Cephaelis ipecacuanha
From 1841, the Royal Pharmaceutical Society of Great Britain acquired a collection of crude drugs. By 1983, when it was donated to Kew, this collection numbered over 9,000 specimens, many of plant origin.

Wax models of filberts
Filberts *Corylus avellana*
Donated by Mrs Mintorn 1899
EBC 69747
Among the most popular exhibits in the original Museum No. 1 were the exquisite wax models of a wide variety of fruits, vegetables and flowers. These were created by members of the Mintorn family.

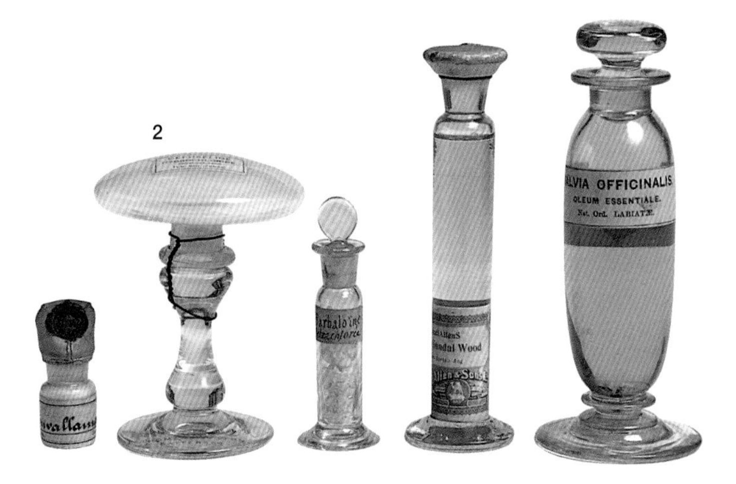

3

Fungi

1 Porcelain model of fungus (3)
Cep Boletus edulis 1888
EBC 69716 Germany

2 Porcelain model of fungus
Gyromitra gigas 1888
EBC 30399 Germany
Kew holds the world's richest and most comprehensive collection of preserved fungi, consisting of over 600,000 specimens. These provide the basis for Kew's continuing research into tropical and British species. These accurately crafted models of fungi were first shown in Museum No. 1.

Plants from Northeast Brazil

1 Soap
Ziziphus joazeiro
Collected by K Pipe-Wolferstan 1996
EBC 73775 Brazil

2 Toothpaste
Ziziphus joazeiro
Collected by R Harley 1997
EBC 73884 Brazil
These objects contain an extract of *Ziziphus joazeiro*, a medicinal plant from Northeast Brazil. Plantas do Nordeste, a joint Kew/Brazilian initiative based in Kew's Centre for Economic Botany, studies the plant resources of this semi-arid area.

Ingredients of potpourri

1 Lili flowers
Fruits of *Soymida febrifuga*

2 Bleached betel halves
Areca catechu
As part of its enquiry service, Kew's Centre for Economic Botany identifies the plants used in various potpourris, checking that they are not toxic or collected from endangered species.

Caring for the Collections

Earrings
Job's tears *Coix lacryma-jobi*
Donated by India Office 1898
EBC 31802 India
Some objects in the Collections are very fragile, in particular the fabrics. Students from the Textile Conservation Centre based at Hampton Court painstakingly clean and conserve items such as these earrings and the accompanying breast plate.

One of the projects supported by Plantas do Nordeste is a medicinal plants nursery in a slum of the city of Fortaleza.

A growing collection

Wooden crocodile bottle opener
Camwood *Pterocarpus soyauxii*
Collected by N Rumball 1995
EBC 60671 Cameroon
New items are constantly being added to the Collections. A member of staff from the Centre for Economic Botany visited Cameroon to collect plant products in everyday use. Each specimen was fully documented with details of the source plant and its use.

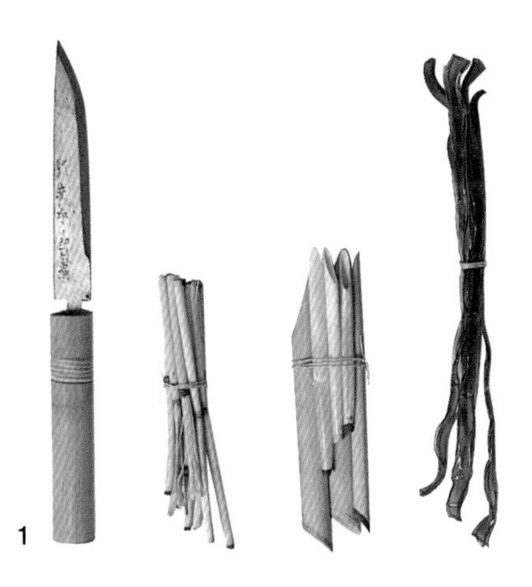
1

Central cabinet 1

Exhibits selected by Kew staff

Inro medicine boxes, lacquerware tools (1) and specimens of lacquer (2)
Lacquer is made from resin produced by various species of *Rhus*, principally *R. verniciflua* which grows in China and Japan. Up to 300 layers of lacquer are applied to wooden bases. Sometimes individual layers are coloured with pigments or incorporate designs created in gold leaf or rice-paper. Specialised lacquerware tools include brushes of rat's or deer's hair, glue made from horsetail stems (*Equisetum* sp.) and bamboo-handled knives.

★ John Quin's lacquer collections were destined for this very Museum, from the moment he acquired them. His lengthy report detailed the cultivation and harvesting of the lacquer trees, the different woods and kinds of lacquer used, prices, production and a listing of what he sent. From the 1880s we get a glimpse of Japanese life, a view of the commercial role of Kew and, of course, three beautiful medicine boxes. HEW PRENDERGAST, *Centre for Economic Botany, Herbanium*

Tools for making and taking snuff
Anadenanthera sp.
The Guahibo people of Venezuela make 'Niopo' snuff from *Anadenanthera* seed ground to a powder in a wooden dish using a pestle made from hard pao d'arco wood (*Tecoma* sp.). They inhale the snuff through the Y-shaped apparatus formed from two hollow bird's bones, each tipped with black balls made from *Astrocaryum* palm fruits. The snuff is stored in a bone snuff box, decorated with aromatic sedges which act as protective charms.

★ As a scientist studying seed chemicals of legumes (members of the pea family), my curiosity was aroused by this snuff-set belonging to the Guahibo people, who make snuff from seeds of the legume *Anadenanthera peregrina*. Their seed-grinding apparatus resembles the equipment that I use to prepare seed – in fact, the wooden dish with its steadying handle is a great improvement over my modern porcelain handle-less mortar! However, the Guahibo inhale powdered *Anadenanthera* seed as a narcotic, whereas I subject ground seed to other forms of chemical analysis! GEOFF KITE, *Biological Interactions Section*, *Jodrell Laboratory*

Carved block of dammar
Agathis dammara
The Malay word 'dammar' means 'resin'. Many conifers, including *Agathis dammara*, the source of this dammar, produce resins when wounded. These resins are often used in varnishes and lacquers because they are insoluble in water and are generally hard and more or less translucent. At one time, in parts of Indonesia and Malaysia, lumps of dammar were softened or ground up and rolled into cylinders to burn as torches.

★ The dammar, a dull brown lump of resin, held no obvious appeal. However, against the light, an elegant carved figure emerged – a glowing amber-coloured object of simple beauty. Its timeless spiritual quality drew me to it. We know that it was carved in the Dutch East Indies over a hundred years ago, but little else. The mystery of who the figure was, and what it may have symbolised, continues to fascinate me. LAURA GIUFFRIDA, *Education*

Cannibal fork and dish
Intsia sp.
These implements were collected independently on the island of Fiji. The fork was donated to Kew in 1861 by Dr Seeman and the plate by Mr Brant in 1880. Made from merbau, the timber produced from *Intsia* species native to Southeast Asia and the Pacific area, these objects illustrate the versatility of this rich heavy wood.

★ This extraordinary cannibal fork and dish always bring a wry smile to my face whenever I handle them, as I remember the squeals of horror (and obvious delight!) from school children and adults alike when shown the items. The one question, however, I'd love to know the answer to is: 'How did the people who donated these items to the collections manage to acquire them without succumbing to the inherent dangers?' GAIL BROMLEY, *Education*

2

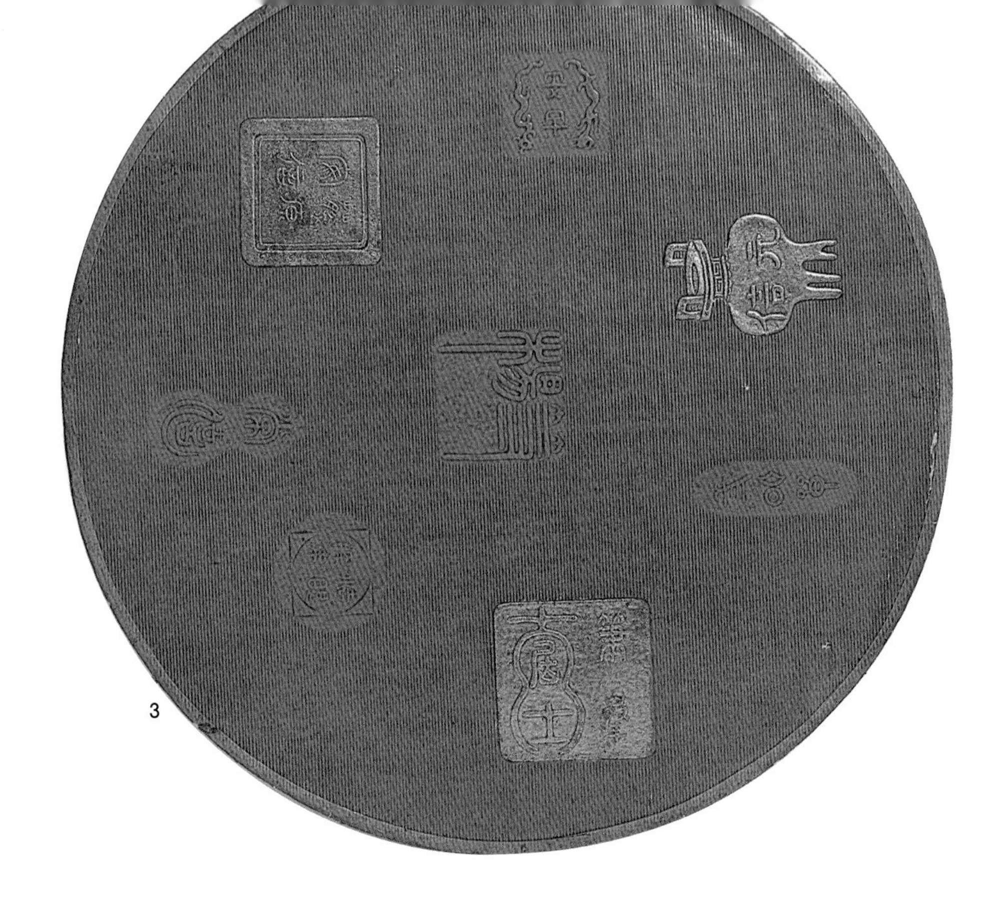

3

Paper seats for summer – paper mulberry (3)
Broussonetia sp.
These seats are made of paper produced from the bark of a paper mulberry (*Broussonetia* sp.) They were sent back to Kew by James Veitch, one of the famous family of nursery owners, who toured Japan in 1893 in search of choice and unusual plants for introduction to European gardens. In his book *A Traveller's Notes* he described the vegetation he encountered and commented on the diverse uses of plants he saw.

★ Is it a plate? Is it a weapon? Is it an ornament? No, it's a paper seat! These seats belong to the astonishing range of products made from paper mulberry (*Broussonetia* sp.) held in the Economic Botany Collections. They remind me of the thrill of opening the storage racks which hold the Collections to discover yet more intriguing objects – some simple, others intricate, often beautiful – and all with a story to tell. PAT GRIGGS, *Education*

Exchequer tally – willow
Salix sp.
This exchequer tally, made of willow (*Salix* sp), represented the sum of £100,000 in part repayment of a loan of £1.4 million pounds from government and to cover the interest of some £6050 due on 30 September 1776. The amount of debt is shown by the number of notches. To record transactions, tallies were split into two unequal pieces that were given to the borrower and the lender.

★ To me, this tally stick embodies man's ingenuity in devising a simple system for recording complex transactions for an age of widespread illiteracy. It also appeals to my sense of humour that our Exchequer was still using this system until 1826 when it had long ceased to be used elsewhere. Some cynics among the staff would, of course, suggest that my love of this object derives from an inordinate love of money and all things associated with it! BILL WEBB, *Administration*

Gramophone needles
Opuntia sp.
These gramophone needles are formed from the spines of the opuntia cactus. Opuntias are native to the Americas, but have been widely introduced to other areas including South Africa, the Canary Islands and Australia. In some places they have spread so widely that they have become a weed. Some species provide edible fruit (prickly pears) or flattened stems (the nopals used in Mexican cooking); others act as extremely effective living fences.

★ Since discovering our Collections' organic gramophone needles, I have run trials on an old record player with a record of the Foden Prize Brass Band. Old needles do not provide the quality professed by the many testimonials in our archives; there is certainly a mellow note but almost inaudible. Newly harvested spines from our Living Collections Department produce similiar results; perhaps more research is required, but they certainly worked in the 1920s! BILL LOADER, *Computer Section, Information Services Department*

Wax model of *Rafflesia* flower
In Victorian times, wax models of plants were very popular. For many years, this giant *Rafflesia* flower greeted visitors as they entered the middle floor of Museum No. 1. Discovered in the Sumatran highlands in 1818 by Sir Stamford Raffles and Dr Joseph Arnold, *Rafflesia arnoldii* grows as a parasite on *Tetrastigma* vines. The malodorous flowers are pollinated by carrion flies and at one metre across are the largest in the plant kingdom.

★ My first visit to Kew was thirty years ago as a young child. Whilst I have only vague recollections of the day, I distinctly remember being lifted up to peer into a large case containing this flower of nightmarish appearance and proportions. Little did I know that some twenty years later, I would try and culture seeds of *Rafflesia* in the Micropropagation Unit at Kew. GREG REDWOOD, *Gardens Development Unit, Living Collections Department*

1

Growing collections 2

Kew's Economic Botany Collections are still growing. New objects are donated and collected to illustrate our continuing dependence on plants. In this way, the Collections keep up to date with the diverse range of plant uses around the world.

Bag (1)
Pandanus spiralis
Collected by S Edwards 1996
EBC 73773 Australia
This bag consists of threads dyed with the roots of *Pogonolobus reticulatus*. It was made by Molly Yawalminy, a Naniyu Aborigine from the Northern Territory, and was purchased in Darwin by a Kew staff member on sabbatical in Australia.

Fruit and carved 'mushrooms'
Raphia taedigera
Donated by A Binfield 1993
EBC 73887
Some specimens reach the collections as a result of enquiries. A wood turner sent in this fruit and the carved ornaments for identification. It proved to be the fruit of a South American palm.

Sieve (2)
Pennisetum sphacelatum
Collected by S Bidgood 1996
EBC 73794 Ethiopia
A botanist from Kew's Herbarium purchased this sieve, made from local plant materials, from a market at Jimma in Ethiopia during an expedition to study the country's flora.

Fibre belt or headband
Allo *Girardinia diversifolia*
Donated by S Dunsmore 1997
EBC 73896 Nepal
The Rai women from the Sankhuwasabha mountains of Nepal extract fibre from a type of nettle and weave it into fabric. This specimen was donated by a British weaver who has studied the use of allo fibre in Nepal.

Large open box
Birch *Betula* sp.
Donated by D A Petelin 1996
EBC 73656 Russia
This box, collected from a tribe in the Russian Far East by a Russian botanist, was donated to Kew staff running a course on herbarium techniques in St Petersburg.

Bark
Virola theiodora
Donated by G T Prance 1996
EBC 73652 Brazil
The Yanomami people of northern Brazil use an exudate of virola bark to prepare a hallucinogenic snuff. This bark was brought back by a former Director of Kew, who has undertaken extensive research in Amazonia.

This picture of Yanomami men was taken by a former Director of Kew in northern Brazil.

2

3 Fabulous fabrics

For warmth in winter, shelter from the summer sun and protection from wind and rain, clothes are a basic necessity. We extract fibres from leaves, seeds or stems and spin them into threads to weave into cloth. Strips of bark are beaten to create other materials, while some so-called 'man-made' fabrics start out as wood.

Clothes from fig trees

Artists have often depicted biblical figures clad in fig leaves, but tree bark is also a source of a clothing material. This apron is woven from bark strips, while the mask is made of beaten bark.

Native apron
Fig *Ficus* sp.
Donated by Sanderson 1859
EBC 43288 South Africa

Tikuna Indian mask (1)
Fig *Ficus* sp.
Donated by G T Prance 1991
EBC 71869 Brazil

1

New Zealand flax

The sword-like leaves of New Zealand flax are the source of a long, lustrous fibre. Maoris used it to make extremely fine, linen-like cloth but it is difficult to extract the fibre mechanically.

New Zealand flax fibre
New Zealand flax *Phormium tenax*
Donated by Carrick, Frost and Co. 1869
EBC 30018

Scotch twilled sheeting
New Zealand flax *Phormium tenax*
Donated by C Thorn 1873
EBC 30010 Great Britain

Ramie

Ramie stems and fibre
Ramie *Boehmeria nivea* var. *tenacissima*
Donated by Consul Wooldridge 1888
EBC 43616 Spain
Ramie fibres are extracted from the stems of a plant closely related to stinging nettles. The thread is very strong, and gets stronger when wet. It is the most silky of all plant fibres.

'Grass' cloth
Ramie *Boehmeria nivea*
Donated by Japan-British Exhibition 1910
EBC 43656 Taiwan

Women artisans preparing tapa cloth in Vatulele, the source of the highest quality material in Fiji. Apart from continuing to be used in local ceremonies, tapa cloth is sold to tourists.

Tapa cloth

Strips of tapa – the paper mulberry's fibrous inner bark – are beaten into sheets which are worn wrapped round the body. King Josiah Tubo's tapa cloth comes from a sheet reputed to be 3.2 km long and 40 m wide.

Two stages in tapa cloth preparation

Paper mulberry *Broussonetia* sp.
Donated by Milne
EBC 42882 Fiji

King Josiah Tubo of Tonga's tapa cloth
Paper mulberry *Broussonetia papyrifera*
Donated by E Howe 1878
EBC 43023 Tonga

Bark cloth from multi-purpose trees

Lace handkerchief
Manila hemp *Musa textilis*
Donated by Dr Veitch
EBC 73132 India
Manila hemp is a relative of the banana plant. Fibres extracted from its tightly-rolled leaf sheaths are used in making paper, rope and sacking, but can also be spun into fine thread.

Lace bonnet
Korean japonica *Chaenomeles lagenaria*
Donated by Lady Doneraile 1855
EBC 57097 Ireland
Lace has been made from flax fibres in Ireland since the seventeenth century, and became particularly fashionable in the mid-1850s. Some makers experimented with other 'botanical fibres'.

Collar made from milkweed fibre (2)
Milkweed *Asclepias curassavica*
Donated by Mrs Cheyne
EBC 73211 Jamaica
Milkweed fruits contain soft, silky hairs. These are extremely difficult to spin because they do not twist around one another, but occasionally a very delicate thread has been produced from them.

2

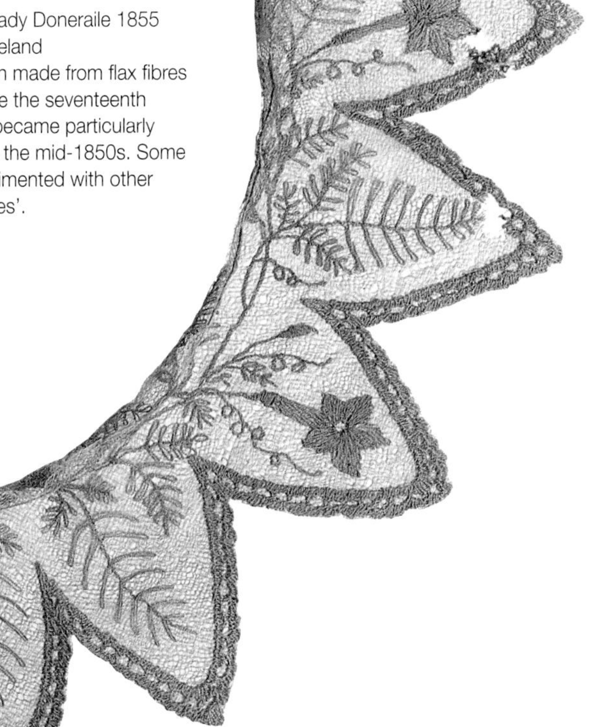

3

4

Linen

Flax stems contain bundles of fibres just beneath the surface. To extract them, the stems are allowed to rot slightly and the waste material is cleaned away. The fibres are combed to separate out the longest for linen thread production.

Dyed flax twine (3)
Flax *Linum usitatissimum*
Donated by Barrow Flax and Jute Co.
EBC 73277 Great Britain

Bundle of flax stems
Flax *Linum* sp.
EBC 64602

Cool cotton

Cotton is the world's most important plant fibre – over 18 million tonnes are harvested each year. We have been making fabrics from it for over 5,000 years, from delicate muslins to tough denim. Cotton thread is made from soft fibres attached to the cotton seeds. Each kilogram of cotton thread contains 200 million seed hairs! Vibrantly dyed cotton is often woven into fabrics with distinctive local patterns. Over 140 million people worldwide depend on cotton picking or handling for their livelihoods.

Cotton plant
Cotton *Gossypium* sp.
Donated by US National Museum 1883
EBC 65598 USA
Cotton fruits, or 'bolls', are filled with seeds surrounded by a mass of seed hairs. One seed can carry 7000 seed hairs, each a single hollow tubular cell. The hairs may be up to 5 cm long.

Hand-spun cotton
Cotton *Gossypium* sp.
Collected by S Bidgood and Melaku 1995
EBC 73613 Ethiopia
1 Clean cotton
2 Second stage twisting
3 Third stage twisting
4 Spindle made from gourd and bamboo
Cotton is still spun by hand in many parts of the world. The wall structure of the cotton hairs causes them to twist as they dry so that they cling to one another when spun into thread.

Javanese body cloth (4)
Cotton *Gossypium arboreum*
Donated by J Henshall
EBC 73230
Gossypium arboreum originated in India and textiles made from the seed hairs of this species date back over 5,000 years. Today, *G. hirsutum* from Central and South America is the most important source of textile fibres.

Finished and carded cotton
Cotton *Gossypium* sp.
Donated by Great Exhibition 1851
EBC 71891
Once the cotton bolls have been harvested, the fibres they contain are cut or pulled away from the seeds. They are cleaned and carded so that the fibres all lie parallel to one another ready for spinning.

Guayacu
Cotton *Gossypium* sp.
Collected by R Spruce 1855
EBC 65600 Venezuela
This guayacu was woven by the Piaroa people. They wore it by passing it between their legs and securing it at the waist. Its free end was draped over the left shoulder or left hanging down behind.

Getting in a lather 4

Plants have long played an important role in our quest for personal hygiene. They feature in many soaps, shampoos and detergents, as well as in flannels, loofahs and cellulose sponges. Toothpastes, toothbrushes and toothpicks also contain some surprising plant products. The earliest toothbrushes were probably chewsticks, which are still used in many parts of the world.

1

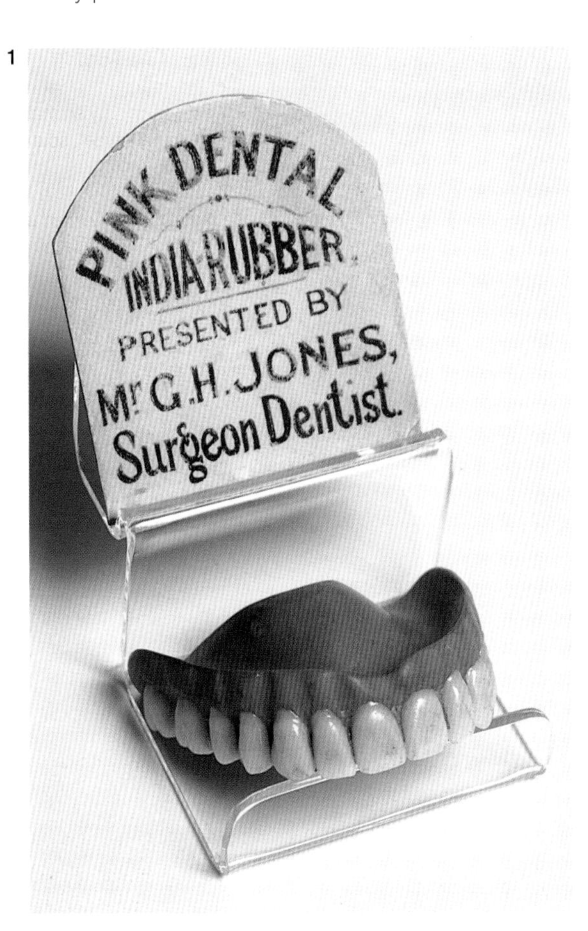

Rubber denture (1)
Rubber *Hevea* sp.
Donated by G H Jones
EBC 44134

Wooden toothbrushes
Viburnum opulus
Donated by Edinburgh Forestry Exhibition 1884
EBC 54025 Japan

Chewsticks
Garcinia mannii
Collected by N Rumball 1995
EBC 73525 Cameroon
In Cameroon, people chew twigs of *G. mannii* to clean their teeth. The twigs are thought to have antibacterial properties. They have been over-collected, and the source tree is rapidly disappearing.

Chewsticks
Bitter kola nut *Garcinia kola*
Donated by International Rubber Exhibition 1914
EBC 66620 Nigeria
Chewsticks made from bitter kola nut twigs are used throughout West Africa as they are considered to be superior to any other wood. They are believed to whiten the teeth and prevent decay.

Toothpastes

Modern toothpastes incorporate many plant products: flavourings such as peppermint, cassia oil, lemon and fennel; carrageenan from seaweed and cellulose gum stabilise the paste; and palm oil glycerine makes it smooth.

Toothpaste
Donated by The Body Shop 1997

Sarakan toothpaste
Toothbrush tree *Salvadora persica*
Donated by F Cook 1988
EBC 50303
This toothpaste contains an extract of *S. persica*, a tree used for centuries by people in India, Sri Lanka and the Middle East for toothbrushes. The tree contains chemicals that prevent decay and gum disease.

Betel nut toothpaste
Betel nut palm *Areca catechu*
EBC 34988
The label on this toothpaste, prepared to an Indian recipe by a London chemist, states that it is for 'cleansing, preserving and beautifying the teeth and gums'. The paste contains ground, dried betel nuts.

Toothpicks

Rays of flowerhead used as toothpicks
Ammi visnaga
Donated by G E Wickens 1987
EBC 1 Egypt
People have found many different plant parts suitable for use as toothpicks – not only strips of wood but also the stiff rays of various flowering heads. The wooden toothpicks are sometimes beautifully carved.

Japanese toothpicks in wooden case
Japanese cedar *Cryptomeria japonica*
1897
EBC 28593 Japan

Toothpicks
Willow *Salix* sp.
Donated by J Henriques 1879
EBC 41399 Portugal

Ylang-ylang flower and foliage, Madagascar.

Smelling sweet

Soaps are made from modified plant oils which not only froth up in water but also attract greasy dirt particles. Palm oil is the most commonly used and is also the main ingredient in margarine, but other oils used include olive, coconut and almond. Fragrant essential oils from plants also scent soaps. In eastern countries ylang-ylang and sandalwood are preferred, while rose and lavender remain popular in European countries. Ylang-ylang flowers look insignificant, but they give a powerful scent.

Soaps

Soaps

Palm oil *Elaeis guineensis*
Donated by The Body Shop 1997
Oil palm fruits contribute palm oil, from the fruit pulp, and palm kernel oil, from the seeds. The oils were initially used for soap and candles, but in the twentieth century became a major constituent of margarines.

Coconut oil soap

Coconut palm *Cocos nucifera*
Donated by The Body Shop 1997
Coconut oil comes from the white flesh just inside the nut. When dried, this flesh is called copra. It is an important item of trade for many Pacific island communities.

Lavender soap

Lavender *Lavandula* sp.
Donated by The Body Shop 1997
Grown for its scent since Tudor times, lavender is one of the most popular fragrances for soaps and other toiletries in western countries.

Sandalwood soap

Sandalwood *Santalum album*
Donated by F N Howes 1928
EBC 44713 India
The heavy, sweet scent of sandalwood is probably the most widely used of all scents for toiletries and perfumes. The fragrant oil is distilled from the wood of small semi-parasitic tropical trees.

Olive oil soap

Olive *Olea europaea*
1997
The cultivation of olive trees around the Mediterranean dates back to the Bronze Age. Oil pressed from the fruits has been used for cooking and lighting and, later, for soap manufacture.

Soap (2)

Chaulmoogra *Hydnocarpus kurzii*
Donated by Department of Commercial Intelligence 1927
EBC 66970 India
Ancient Indian religious texts detail the use of chaulmoogra oil to treat leprosy and skin complaints. Its source was not known in Europe until the 1920s when the seeds of *Hydnocarpus kurzii* were identified.

Soapnuts and powder

Soapnut *Sapindus mukorossi*
Donated by Global Eco 1996
EBC 73780 India
Some plants contain chemicals that are naturally soapy and produce foam when mixed with water. Soapnuts are used in India to wash silks and other delicate materials such as Kashmir shawls.

Soap root

Gypsophila sp.
Donated by Professor Fluckiger 1891
EBC 66807 Turkey

2

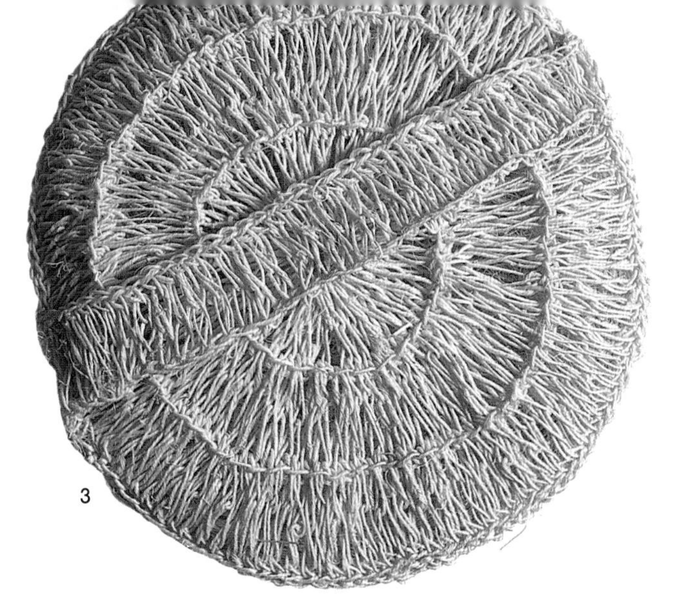

3

Shampoos

Various herbal ingredients are added to shampoos for their beneficial properties. For example, rosemary (*Rosmarinus officinalis*) is believed to impart a sheen and deepen hair colour.

Rosemary and cinchona shampoo
Rosmarinus officinalis
Cinchona sp.
Donated by The Body Shop 1997

Seaweed and peony shampoo
Paeonia sp.
Donated by The Body Shop 1997

Stem used as hairwash
Entada rheedi
Donated by British Empire Exhibition 1925
EBC 59031 Malaysia
The bark and seeds of *Entada* are rich in saponins – chemicals that foam when mixed with water. In Southeast Asia they are used as shampoos and as a cleanser for wounds and lice infestations.

Henna shampoo
Henna *Lawsonia inermis*
Donated by The Body Shop 1997
Although henna has long been used to tint hair red, it is also considered to improve hair quality, giving it shine and increased strength.

Sponge of stem fibres
Acacia sp.
Collected by N Rumball 1995
EBC 73530 Cameroon
Coarse fibres, extracted from the stems of a climbing species of *Acacia*, are used as body scrubs throughout Cameroon. The climber is becoming increasingly rare in the wild due to overharvesting.

Loofah fruit
Sponge gourd *Luffa cylindrica*
Donated by RBG Kew 1889
EBC 54751 Great Britain
Beneath the skin of this long cylindrical gourd lies a complex mesh of cream fibres – recognisable as the loofah used at bathtime.

Bath scrub (3)
Agave spp.
Donated by Nanhu people 1994
EBC 73223 Mexico
This bath scrub is made from fibres extracted from two species of *Agave*. It is being marketed in this country as part of a sustainable development initiative.

Flannels
Cotton *Gossypium* sp.
Donated by The Body Shop 1997
Cotton fibres are particularly useful as flannels because they readily absorb water and their strength increases when they are wet.

2

5 Healing plants

Over three quarters of the world's population depend directly on plants for medicine. Many traditional medical systems rely almost entirely on plant materials. Even in western medicine, one quarter of all drugs use active ingredients from natural sources. The plant kingdom has great potential as a source of new drugs. As recently as 1994, taxol – a drug produced from yew – became available to treat breast cancer.

Decongestants from ephedra

Soma plant
Ephedra sp.
Donated by Delimitation Commission 1886
EBC 27165 Afghanistan

Decongestant linctus
Ephedra sp.
1985
EBC 55205
Traditional Chinese doctors prescribe *Ephedra* to treat asthma and chills. As Chinese remedies have become popular in the West, Kew has set up a project to authenticate the plant materials used.

Arrow poisons and anaesthetics

Amazonian people use various plants in their arrow poisons. One, the *Chondrodendron* vine, contains a chemical that paralyses animals. Anaesthetists use this chemical, tubocurarine, to relax a patient's muscles.

Yanomami bamboo quiver (1)
Donated by G T Prance 1992
EBC 71824 Brazil

Roots
Chondrodendron tomentosum
Donated by Royal Pharmaceutical Society 1983
EBC 67446

Tubocurarine injection
Chondrodendron sp.
Donated by Wellcome Foundation Ltd 1984
EBC 55204

1

Medicinal plant market, Anguo, Hebei, China.

2

4

A bark to treat malaria

Malaria has caused more deaths than any other disease in history. From the seventeenth century, cinchona bark and, later, the extracted quinine were the best treatments for malaria until artificial drugs were developed. But malarial parasites are developing resistance to these, and so doctors are starting to return to natural quinine. The cinchona tree's bark yields quinine. The name 'quinine' comes from the South American tree's name 'quina quina' – bark of barks. Kew scientists are investigating the chemicals from various plants as potential antimalarial drugs.

In the mid-seventeenth century, Jesuit missionaries in South America observed local people treating fevers with cinchona bark. Sent back to Europe, the bark became the first successful treatment for malaria.

Bark
Cinchona succirubra
Donated by Royal Pharmaceutical Society 1983
EBC 56225
a natural stem bark
b natural branch bark
c bark from C. *succirubra var. subpubescens*
d Jamaica red bark
e *Cinchona* hybrid stem bark

Ordinary cinchona febrifuge
Cinchona *Cinchona pubescens*
Donated by G King 1882
EBC 52455 India
As the number of wild cinchona trees in South America declined, the British and Dutch sought to establish plantations in the Far East. Seeds were sent back to Kew in 1860, and from there to India and Sri Lanka.

Packets of quinine (2)
Cinchona *Cinchona pubescens*
Donated by J S Gamble 1925
EBC 52445 India
Malaria is a major killer in India. These packets, each containing 5 grains of pure quinine, were sold at post offices. Many Europeans took their daily dose in the form of refreshing tonic water.

Branch bark
Cinchona *Cinchona officinalis*
Donated by India Museum 1880
EBC 52643 India

Quinine bisulphate tablets
Cinchona *Cinchona* sp.
EBC 55198
Natural quinine is still effective against malaria, although it has largely been replaced by synthetic forms of the drug.

Donations from the Royal Pharmaceutical Society

Materia Medica collection (3)
Donated by Royal Pharmaceutical Society 1983
EBC 73047
Until the mid-nineteenth century, doctors relied heavily on plant drugs. Plants still provide various important drugs and the raw material needed to synthesise others. This collection consists of 111 samples of medicinal plants.

Willow bark (4)
White willow *Salix alba*
Donated by Royal Pharmaceutical Society 1983
EBC 42212
Willow leaves and bark have been used for centuries to treat aches and pains. In 1827, the active ingredient, salicin, was isolated. Its name came from the willow's scientific name – *Salix*.

3

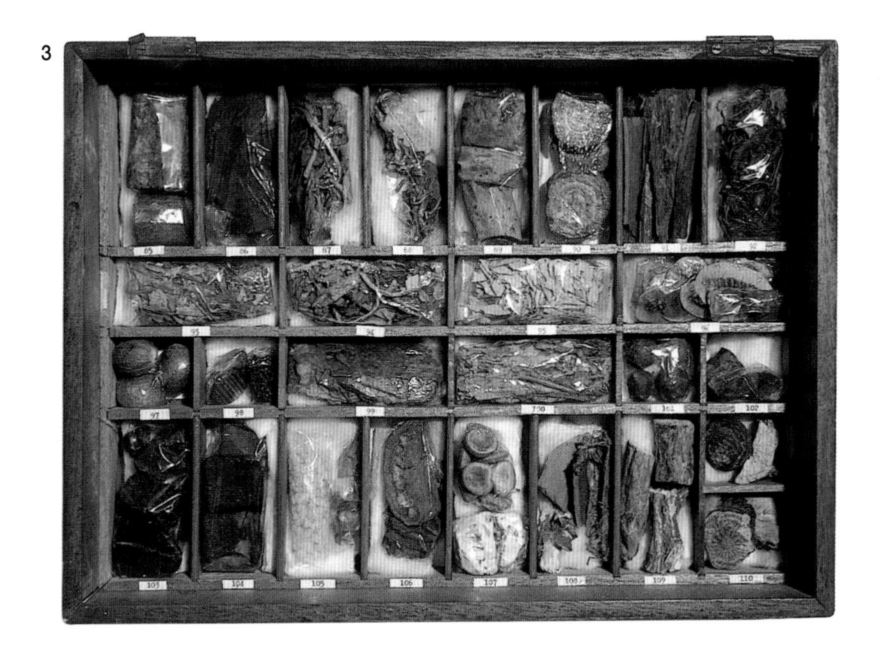

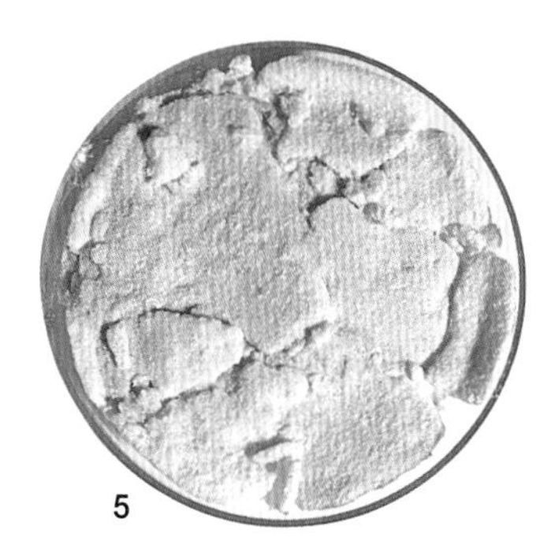

5

Cameroonian botanist with trunk of a *Prunus africana* tree stripped of its bark.

Salicin crystals (5)
Willow *Salix* sp.
Donated by Royal Pharmaceutical Society 1983
EBC 71995
During the nineteenth century, salicin from willow and the related salicylic acid from meadowsweet were both used for pain relief. A laboratory derivative of the acid is a familiar medicine – aspirin.

Foxgloves, Great Britain.

Heart drugs from foxgloves

In the eighteenth century a British doctor noted that a woman had cured herself of dropsy by taking foxglove. Its heart-regulating chemicals were identified over 60 years ago, but have never been made artificially.

Foxglove leaves
Foxglove *Digitalis purpurea*
Donated by Bowyer and Bartleet 1920
EBC 47872 Japan

Digoxin tablets
Foxglove *Digitalis purpurea*
Donated by Wellcome Museum of Medical Science 1984
EBC 55197

Madagascar periwinkle

Leaves
Madagascar periwinkle *Catharanthus roseus*
Donated by Royal Pharmaceutical Society 1983
EBC 50192 South Africa
The Madagascar periwinkle has been used against diabetes for centuries. It is now known for treating various forms of cancer, including children's leukaemia. 53 tonnes of leaves provide just 100 grammes of the drug.

Vincristine injection
Madagascar periwinkle *Catharanthus roseus*
Donated by Bristol Myers Pharmaceuticals
EBC 55202

Snakeroot

Root
Snakeroot *Rauvolfia serpentina*
Donated by Royal Pharmaceutical Society 1983
EBC 50681
For over 4000 years, Indian healers have used snakeroot against snakebites and mental illness. The root contains chemicals that relieve high blood pressure. One chemical, reserpine, is also used to treat anxiety and schizophrenia.

Tablets for high blood pressure
Snakeroot *Rauvolfia serpentina*
EBC 55203

African cherry

Cherry Bark
African cherry *Prunus africana*
Collected by N Rumball and A McRobb 1995
EBC 73557 Cameroon
Extract of African cherry bark is an effective treatment for prostate cancer. The bark is collected from wild trees found in several African countries. Overharvesting is threatening the survival of some populations of the species.

Highs and lows 6

Some plant extracts can alter how we feel, making us relaxed, alert or dreamy. Many cultures have used such extracts for pain relief and anaesthesia, and sometimes for ceremonial purposes or pleasure. Many are now illegal unless prescribed, because they are addictive and poisonous.

1

Betel nuts

Betel nut basket made of split bamboo
Bamboo
Donated by Consul Swinhoe
EBC 40629 Taiwan
In many parts of Asia and Polynesia, people chew betel nuts, especially during celebrations and ceremonies. Betel chewing causes continuous salivation and turns the saliva bright red.

Betel nuts
Betel nut palm *Areca catechu*
Donated by Royal Pharmaceutical Society 1983
EBC 56331
Betel nuts are part of the seeds of the betel nut palm. People chew slices of nut wrapped in a leaf of betel pepper (*Piper betle*) for their narcotic effect. It is also believed to strengthen teeth and sweeten breath.

Flavourings for betel nut quids (1)
Betel box made from cloves
Clove *Syzygium aromaticum*
Donated by J Henshall
EBC 55640 Indonesia

Chunam box to hold lime
Bottle gourd *Lagenaria siceraria*
Donated by HMS Challenger Expedition 1875
EBC 54651 New Guinea
A wide range of additives are incorporated into the betel nut and betel pepper 'quids'. Two of the most popular are cloves and powdered lime. Other additives include tobacco and cardamoms.

Kava

Kava roots
Kava *Piper methysticum*
Donated by Colonial and India Exhibition 1886
EBC 45529 Fiji
Kava is the national beverage of the Polynesian islanders and is closely linked to their social, religious and political life. It is made from the roots of a type of pepper, and has sedative and hypnotic properties.

Kava bowl
Intsia bijuga
Donated by S W Silver 1878
EBC 59802 Pacific Islands
At one time, people with strong teeth chewed kava roots and spat their saliva into kava bowls to make the drink. Today, the roots are usually pulped and fermented in water.

2

Powerful poppies

Opium poppy capsules produce a white latex that contains many psychoactive chemicals, including morphine and codeine. Opium and morphine were once widely used for pleasure, but both proved addictive, as did heroin, a derivative of morphine. However, some chemicals from opium are useful for their painkilling effect. Though opium is grown as a cash crop in some countries, only four per cent is harvested for legitimate purposes. Slashing the poppy capsule releases the potent latex.

Opium poppy capsules (2)
Opium poppy *Papaver somniferum*
Donated by Royal Pharmaceutical Society 1983
EBC 67406 Turkey
Latex from opium poppies contains 26 active chemicals. Two are often used medicinally; morphine is addictive and is used only as a potent pain-killer for serious illness, while milder codeine is not habit-forming.

Capsules and opium collecting tools
Opium poppy *Papaver somniferum*
Donated by Royal Pharmaceutical Society 1983
EBC 71993 Turkey and Yugoslavia
Harvesters slash the capsules so that the white latex oozes out. When the latex has dried to a tacky mass, it is scraped off and dried further. Opium poppy seeds, used for baking, do not contain any active chemicals.

Apparatus for opium smoking (3)
Donated by C Ford 1881
EBC 41313 Hong Kong

Opium pipes
Bamboo
T Watters 1888
EBC 41322
Opium was originally taken dissolved in wine, but by the eighteenth century it was more often smoked. A heated ball of latex was dropped into a pipe and the smoker inhaled the vapours given off.

Models of opium smokers
Euonymus sp.
Donated by W M Cooper 1879
EBC 41297 China
Opium smoking became widespread in China during the eighteenth century and British companies imported the drug to sell. The Opium Wars (1839–41) resulted from official Chinese opposition to this trade.

3

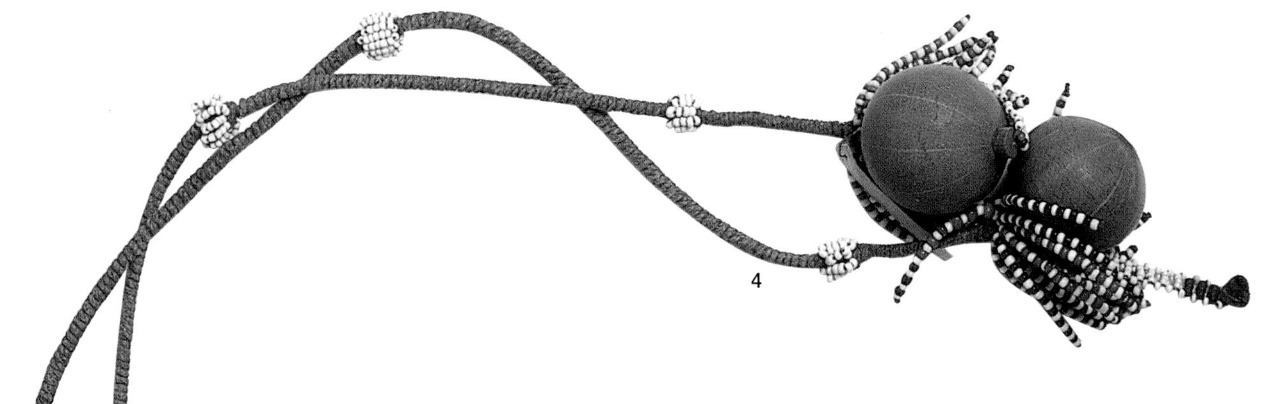

Tobacco

Tobacco originated in Central and South America, but is now grown in many warm countries. China and the USA are the major exporters, but many other Asian and African countries produce tobacco.

Tobacco leaves on a stick
Tobacco *Nicotiana* sp.
EBC 48369 Burma

Roll of native tobacco
Tobacco *Nicotiana* sp.
Donated by Niger Expedition 1859
EBC 48428 Nigeria

Pigtail tobacco
Tobacco *Nicotiana* sp.
Donated by Lambert and Butler 1885
EBC 48194
Twist tobacco is used for both chewing and smoking. It is made from darkly coloured, strongly flavoured tobacco.

Sailors' plug tobacco
Tobacco *Nicotiana* sp.
1896
EBC 48381
Plug tobacco, formed from dark, air- and fire-cured tobacco leaves, is chewed rather than smoked. The first plug tobaccos were soaked in honey and plugged into holes in hickory or maple logs – hence the name.

Flue-cured tobacco
Tobacco *Nicotiana* sp.
Donated by Bishop of Rhodesia 1956
EBC 48396 South Africa
Once tobacco leaves have been harvested, complex chemical changes begin. Curing is designed to control the speed at which these occur. Mild tobaccos are cured in heated flues while dark tobaccos are cured in air or smoke.

Cigarettes

Cigarette smoking today accounts for over 80 per cent of British tobacco consumption. Most cigarettes enclose the tobacco in paper, but Indian cigarettes, or bidis, are wrapped in tendu leaves (*Diospyros melanoxylon*).

Russian cigarettes
Tobacco *Nicotiana* sp.
Donated by E Rutenberg 1938
EBC 48297 Russia

Bidi cigarettes
Tobacco *Nicotiana* sp.
Donated by G K E Copeland 1963
EBC 48389 India

Indian cigars
Tobacco *Nicotiana* sp.
EBC 48392 India
Cigars are made from three types of air-cured tobacco leaves – filler, binder and wrapper. Long narrow strips of filler are bunched together with a binder and then rolled spirally in half leaves of wrapper tobacco.

Snuff

When tobacco first reached Europe, people inhaled it as snuff, often to cure headaches. It is made from a mixture of dark air- and fire-cured leaves, finely powdered and mixed with flavourings and scents.

Double Zulu snuff boxes (4)
Snuff box tree *Oncoba spinosa*
Donated by Captain Garden and Mrs Sanderson
EBC 66979 South Africa

Snuff (5)
Tobacco *Nicotiana* sp.
Donated by Lundy, Foot and Co.
EBC 48278, 48279, 48280 Ireland

7 Pick-me-up plants

Coffee from Africa, tea from Asia and chocolate from America are among the world's favourite beverages. They are all important trade commodities, often cultivated far from their place of origin. All three contain caffeine or other gentle stimulants. Some people prefer their drinks without caffeine, but to extract it involves using harmful chemicals. Kew scientists are looking for low-caffeine species of coffee in Madagascar.

Over 180 million cups of tea are drunk each day in Britain alone. Many people take it with milk or lemon, but in China tea is mixed with onions, ginger and orange. Russians add jam, and in Mongolia it is made into soup with flour and salt. Tea is still picked by hand – only the youngest shoot tips will do! In the Japanese tea ceremony, there are several different stages and rituals associated with brewing the green tea. Tea is the world's most popular drink, apart from water.

Teas

Thousands of different types of tea are available around the world – but all are *Camellia sinensis*. They vary in colour, flavour and aroma according to the growing conditions and methods of processing.

Tea compressed into cakes
Tea *Camellia sinensis*
Donated by L Oliphant
EBC 66350

Wheat sheaf tea
Tea *Camellia sinensis*
Donated by Dr Dinwoodie 1808
EBC 66356 China

Superior brick tea (1)
Tea *Camellia sinensis*
Donated by T L Bullock 1892
EBC 66344 China

Fancy tea
Tea *Camellia sinensis*
Donated by Royal Pharmaceutical Society 1983
EBC 66455

Large ball tea
Tea *Camellia sinensis*
Donated by J Reeves
EBC 66362

Gunpowder green tea
Tea *Camellia sinensis*
EBC 68954 China
Most tea produced in China and Japan is green tea. The leaves are steamed or dried immediately after picking to retain their delicate flavour and pale colour. In Japan, green tea is central to the Zen Buddhist tea ceremony

1

Decorated tea brick
Tea *Camellia sinensis*
Donated by B Verdcourt 1996
EBC 45706 China
Brick teas are made from compressed tea leaf fragments and tea dust. They were once an important item of trade between China and Russia. In Russia, chunks were broken off the bricks and infused in a special urn called a samovar.

Tea and tea-colouring samples
Tea *Camellia sinensis*
Donated by B Seeman
EBC 66422 China

Tin of Japanese tea
Tea *Camellia sinensis*
Donated by C Stirton 1992
EBC 71842 Japan

Darjeeling tea
Tea *Camellia sinensis* 1997
Darjeeling is one of the main tea cultivation areas in India, which is the world's largest tea producer. This is a black tea, produced by fermenting the leaves after they have been withered and rolled.

2

Box for rancid yak's butter
Donated by Dr King 1897
EBC 66447 Tibet
Tibetans make their tea from brick tea that is boiled for some hours to produce a thick black brew. They then add salt or soda, strain the liquid, and finally stir in rancid yak butter.

Tea dasher used for churning tea
Donated by Dr King 1897
EBC 66449 Tibet

Bamboo tea strainer
Donated by Dr King 1897
EBC 66436 Tibet

Guarana

Powdered guarana (2)
Guarana *Paullinia cupana*
Collected by H D V Prendergast 1996
EBC 73763 Brazil
Guarana is a vine native to Brazil. Its caffeine-rich seeds are used by Amazonian people to make a stimulating drink, and are the basis of various popular soft drinks. Guarana is now available in Europe.

Guarana rod
Guarana *Paullinia cupana*
Donated by R Spruce 1850
EBC 62373 Brazil
Guarana seeds are removed from the fleshy fruits and then dried and roasted. Finally the seeds are crushed with water and cassava flour to form a paste which is made into rods.

Fish model made from guarana and fish tongue grater
Guarana *Paullinia cupana*
Donated by Royal Pharmaceutical Society 1983
EBC 73099 South America
The rough tongue of an Amazonian fish – the piraruca – is used to grate guarana rods to produce powder. This model of the piraruca is itself made from guarana paste.

Cocoa

Cocoa pods
Cocoa *Theobroma cacao*
EBC 65145
After picking, cocoa pods are cracked open to expose the pulp and seeds. These are scraped out and fermented in the sun, causing chemical changes to take place and the flavour and scent of chocolate to develop.

Cocoa beans
Cocoa *Theobroma cacao*
1928
EBC 65100 West Indies
Fermented and roasted seeds or beans are cracked open to release the 'nibs' inside which are used to make chocolate. The Aztecs of Central America valued the beans so highly that they were used as money – 100 beans would buy a slave.

Criollo cocoa chocolate
Cocoa *Theobroma cacao*
Donated by British Empire Exhibition 1924
EBC 67799 Jamaica
The Latin name for chocolate, *Theobroma*, meaning 'food of the gods', recognises the Aztec belief that cocoa was of divine origin. They drank it mixed with vanilla, chilli pepper and ground annatto seeds.

3

Coffee

Coffee beans are the seeds of certain tropical evergreen shrubs (*Coffea* spp.) native to Ethiopia. The seeds are separated from the pulp of the ripe red fruits and fermented for a day. This process begins the development of the coffee flavour.

Coffee beans
Coffee *Coffea* sp.
Donated by W and G Law 1817
EBC 53570

Mocha coffee beans (3)
Coffee *Coffea arabica*
Donated by Royal Pharmaceutical Society 1983
EBC 53314

Coffee grounds
Coffee *Coffea arabica* and *C. robusta*
Donated by N Rumball and A McRobb 1995
EBC 73555 Cameroon
Two species of coffee are grown commercially. The highest quality coffee beans, used for ground coffee, come from *C. arabica*. *C. robusta* produces poorer quality beans, which are often made into instant coffee.

Coffee leaf tea
Coffee *Coffea* sp.
Donated by J Haddon and Co. 1893
EBC 53583 Sri Lanka
In some places, a tea is made from the leaves of coffee plants. Like the seeds, coffee leaves contain caffeine but at much lower levels.

Maté

Maté is made from the leaves of a tall evergreen tree, *Ilex paraguariensis*, related to holly. After drying and crushing, the caffeine-rich leaves are ready for use. Maté tea is most popular in Argentina.

Powdered maté leaves are placed in a gourd and boiling water is added. After the mixture has brewed, the drinker inserts the straw or 'bombilla' and sucks up the liquid. The holes in the straw's expanded end sieve the drink.

Maté gourds
Donated by Royal Pharmaceutical Society 1983
EBC 63540

Maté tubes
Donated by J T von Allen, F Strangways and J Whiffin
EBC 62867

Real maté tea
Maté *Ilex* sp.
1971
EBC 62873 (Brazil)

Yerba maté
Maté *Ilex paraguariensis*
Donated by S Renvoize 1979
EBC 62863 Argentina

8 Sugar and spice

Pollinating vanilla flowers in Madagascar.

Without herbs and spices for flavourings, and sugar for sweetening, our food would be very bland. Herbs are generally fresh leaves, while spices range from buds to bark and seeds. They often provide the distinctive flavours of regional dishes. The European search for spices in the fifteenth century drove many voyages of exploration. Columbus was seeking a new spice trade route when he reached the Americas in 1492.

1

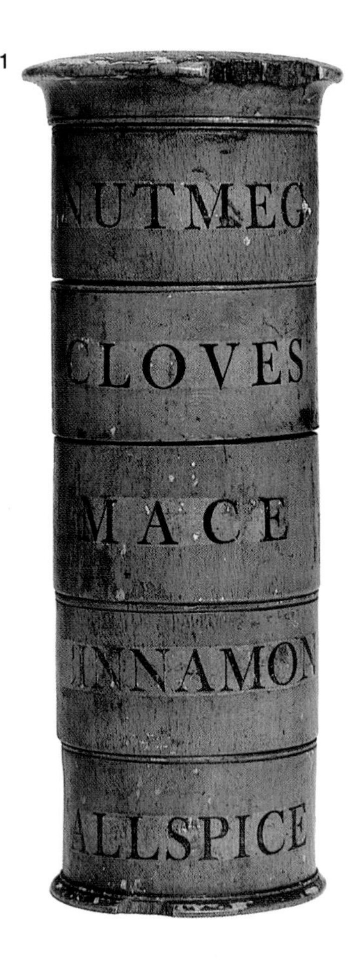

Cinnamon

Cinnamon sticks
Cinnamon *Cinnamomum zeylanicum*
EBC 69750
Cinnamon sticks are strips of the aromatic inner bark of shoots of *C. zeylanicum*, which roll up as they dry. Powdered cinnamon is a common ingredient of cakes and pickles, and it also features in various medicines.

Pepper

Different types of peppercorn have slightly different flavours and scents. Black pepper consists of dried unripe fruits while white pepper comes from riper fruits soaked in running water to remove the surface layers.

Pepper
Piper nigrum
Donated by Royal Pharmaceutical Society 1983
EBC 46889
1 White Singapore pepper
2 Black Sumatra pepper
3 Brown Penang pepper
4 Brown Malabar pepper

Tailed pepper fruits
Tailed pepper *Piper cubeba*
Donated by Royal Pharmaceutical Society 1983
EBC 47155 Java
Tailed pepper, a native of Indonesia, is now only used medicinally in Asian countries, although at one time it was used as a spice in Europe. The spice consists of the dried stalked unripe fruits.

Immature fruits of Indian long pepper
Long pepper *Piper longum*
Donated by Royal Pharmaceutical Society 1983
EBC 46873
Long pepper, from India, was probably the first type to reach Europe. It consists of unripe fruits embedded in a fleshy stalk. In India it is still used in pickles, preserves and curries.

Mandje basket for collecting pepper
Donated by Dr Treub
EBC 45597 Indonesia
Peppercorns are the fruits of a climbing vine from India. The fruits are closely packed together on spikes, which are picked before the corns are separated.

Allspice

The flavour of allspice is described as a combination of clove, cinnamon and pepper. Allspice is one of only three spices native to the New World – the others are vanilla (*Vanilla planifolia*) and chillies (*Capsicum* spp.).

Allspice
Pimenta dioica
EBC 55544 West Indies

Allspice
Pimenta dioica
Donated by Gold Coast Botanic Gardens 1910
EBC 58500 Ghana

Wooden spice container (1)
EBC 72010
This container holds four spices from the wet tropics of Asia and one – allspice – from the New World. The search for new routes to Asia for spices inspired many voyages of exploration during the fifteenth century.

Nutmeg and mace

Nutmeg trees produce two spices – nutmeg from the seed kernels, and mace from the net-like outgrowth that encircles the seed.

Mace
Nutmeg tree *Myristica fragrans*
Donated by J Reeves
EBC 45399 Malaysia

Nutmegs (2)
Nutmeg tree *Myristica fragrans*
EBC 45398 Malaysia

Nutmeg harvester
Donated by H G Forbes 1883
EBC 45459 Indonesia
The apricot-sized fruits of the nutmeg tree are harvested by gripping them with the prongs and pulling so that they drop into the basket.

Vanilla

Vanilla pods are the fruits of an orchid native to Central America. When picked, the pods are green, and they are wilted, sweated, dried and conditioned before the characteristic aroma and flavour develop.

Vanilla pods
Vanilla *Vanilla planifolia*
Donated by Royal Pharmaceutical Society 1983
EBC 37230 Seychelles

Box inlaid with vanilla motif
Madagascan rosewood *Dalbergia* sp.
Donated by P J Cribb and D Du Puy 1994
EBC 47191 Madagascar

Cloves

Cloves are the aromatic, dried flower buds of a small tree native to the Moluccas in eastern Indonesia. They are often added to apple pies and pickles. Oil of cloves is also used in dental treatments.

Cloves
Syzygium aromaticum
Donated by Royal Pharmaceutical Society 1983
EBC 56616 Seychelles

Cloves
Syzygium aromaticum
Donated by India Museum 1880
EBC 55605 Malaysia

Cloves
Syzygium aromaticum
Donated by International Exhibition 1862
EBC 55601 Mauritius

Model of Malay boat made of cloves
Syzygium aromaticum
Donated by J Henshall 1854
EBC 55560 Indonesia

A Chilean wine palm dominates in this view of Kew's Temperate House.

Sweets for my sweet

All green plants make sugars, but most of the world's processed sugar comes from sugar cane and sugar beet. Sugar cane is a giant grass native to New Guinea that is now grown throughout the wet tropics. Until the sixteenth century, sugar was an expensive luxury in Europe – a kilogram bag would have cost £125 in today's terms. Much sugar cane is harvested by hand, using sharp machetes to cut the canes. Sugar beet is a major crop in many temperate countries. Mexico's 'Day of the Dead' celebrations involve skull-shaped sugar sweets.

Sugar cane stem

Sugar cane *Saccharum officinarum*
Donated by Royal Pharmaceutical Society 1983
EBC 33992
Sweet-stemmed sugar canes probably originated in New Guinea. Originally tribespeople chewed them for their sugary juice. Lengths of cane provided a convenient food supply for journeys.

Model of sugar cane mill

Donated by India Museum 1880
EBC 32445 India
Canes are crushed between the rollers and the sweet juice is pressed out. The fibrous residue is used as fuel and to make paper. Modern sugar mills use larger scale mechanised rollers.

Sugars

Sugar cane *Saccharum officinarum*
EBC 32502, 71913, 71914, 71919, 71921
Boiling water off cane juice leaves a thick liquid which is filtered and then spun. This separates the crystals of sugar from the surrounding brown molasses. Brown sugar retains some molasses.

Raw beet sugar

Sugar beet *Beta vulgaris*
EBC 71945
The swollen roots of sugar beets contain up to 20 per cent of a sugar identical to that in sugar cane. Unlike sugar cane, sugar beet grows well in temperate countries.

Maple sugar

Sugar maple *Acer saccharum*
EBC 62223 Canada
The trunks of the sugar maple are tapped in spring to collect the sweet sap which is boiled down to make sugar and syrup. Maple sap contains just 8 per cent sugar compared with the 22 per cent in sugar cane.

Palm honey (3)

Chilean wine palm *Jubaea chilensis*
Donated by S Henchie and A Kirkham 1985
EBC 35739 Chile
Palm honey is produced from the sugary sap of the wine palm. The sap is tapped from the trunk and then boiled down until a thick syrup forms.

3

Taking our pulses 9

Pulses are edible members of the legume family, which includes beans, peas and peanuts. For millions of people throughout the world, especially vegetarians, these plants are a vital source of protein. Growing legumes is also good for soil fertility, because special bacteria living in their roots 'fix' soil nitrogen and make it available for other plants.

1

Lentils

Lentils are among the world's oldest food crops; archaeological excavations in the Middle East have found 8000 year old lentils. They are easily digestible and in India often form the basis of purées and soups.

Lentils
Lentil *Lens culinaris*
Donated by Great Exhibition 1851
EBC 60623 Egypt

Lentils
Lentil *Lens culinaris*
EBC 60609

Split lentils
Lentil *Lens culinaris*
EBC 60629 India

Broad beans

Broad bean seeds
Broad bean *Vicia faba*
Donated by Sutton and Sons
EBC 61413
The broad bean was the only edible bean known in Europe until the end of the fifteenth century, when explorers reached the Americas and discovered the wide range of beans under cultivation there.

Pigeon peas

Pigeon pea is a perennial shrub which can tolerate poor soils, growing on land unsuitable for other crops. The seeds are eaten both fresh and as dried pulses, especially in the West Indies in 'rice and peas'.

Pigeon peas
Pigeon pea *Cajanus cajan*
Donated by India Museum 1880
EBC 60100 India

Pigeon peas
Pigeon pea *Cajanus cajan*
Donated by India Museum 1880
EBC 60119 India

Pigeon peas
Pigeon pea *Cajanus cajan*
Donated by India Museum 1880
EBC 60105

Groundnuts

Groundnuts (1)
Groundnut *Arachis hypogaea*
1926
EBC 59966 The Gambia
Groundnuts, also called peanuts, originated in South America. The groundnut actually plants its own seeds – long stalks bearing the developing fruit push below the soil and the fruit matures underground.

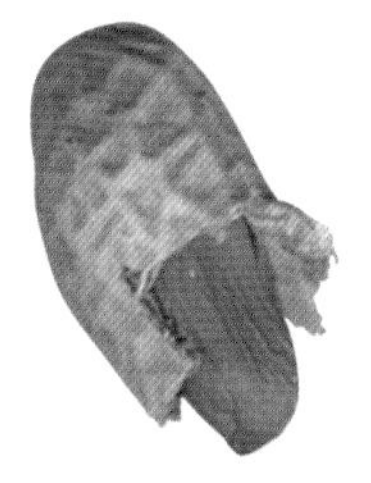

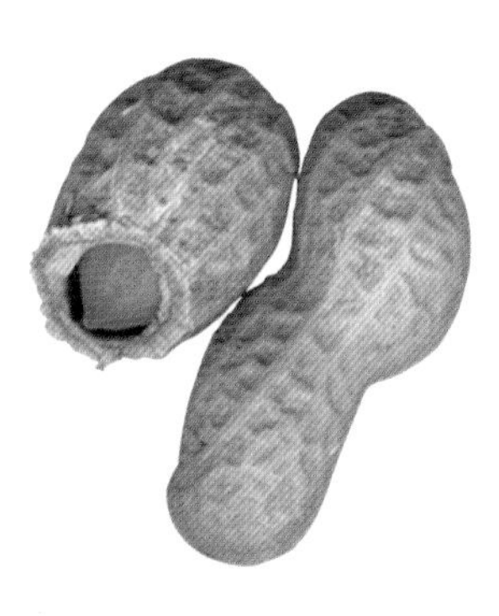

2

Beans

Young pods and the tender beans they contain are popular vegetables in India. The seeds are also dried and eaten as a pulse. Hyacinth bean plants can withstand drought and grow on land too dry for other pulses.

Hyacinth beans
Hyacinth bean *Lablab purpureus*
Donated by T L Bullock 1897
EBC 60318 China

Hyacinth beans
Hyacinth bean *Lablab purpureus*
Donated by India Museum 1880
EBC 60319 India

Varieties of French bean

The fresh young fruits of *Phaseolus vulgaris* are eaten as French or string beans. Its dried seeds are variously known as kidney, haricot and pinto beans. Cooked in tomato sauce, they are 'baked beans'.

French bean *Phaseolus vulgaris*
Donated by Carters 1952
EBC 61064

1. Brown Dutch
2. Melting Dwarf Wax
3. Climbing Perfection
4. 50 Days
5. Merton Haricot
6. July Climbing
7. Climbing Purple Podded
8. Pershore
9. Lighting
10. Sunrise

Chickling vetch

Chickling vetch seed
Chickling vetch *Lathyrus sativus*
Donated by Royal Pharmaceutical Society 1983
EBC 58293 India
Chickling vetch is grown as food for both people and animals. It is the cheapest pulse in India, and is eaten there when food is short. If consumed over a long period, the seeds can cause a paralysing disease.

Tamarind

The pulp of tamarind fruits supplies a fruity, sour flavour to various Asian dishes and is also used to make refreshing drinks and sweets. It is an ingredient of Worcestershire sauce and Angostura bitters.

Tamarind pods
Tamarind *Tamarindus indica*
Donated by Fortnum and Mason Ltd
EBC 73255 West Indies

Block of tamarind pulp
Tamarind *Tamarindus indica* 1997
Thailand

Liquorice

Although once widely cultivated in England, liquorice became restricted to part of Yorkshire, where 'Pontefract cakes' were produced from it. It is now often used medicinally in cough syrups and pastilles.

Liquorice stems
Liquorice *Glycyrrhiza glabra*
Donated by Royal Pharmaceutical Society 1983
EBC 56231

Liquorice sticks
Liquorice *Glycyrrhiza* sp.
1997

Gums

Gum arabic (2)
Gum arabic *Acacia senegal*
EBC 73040
When gum arabic trees are damaged, they exude a gum to seal the wound. This gum, still collected from wild trees in Northeast Africa, features in many products, from food to paint.

Gum tragacanth exuding from stem (3)
Gum tragacanth *Astragalus gummifer*
Donated by J D Hooker
EBC 60040 Palestine
Gum tragacanth bushes exude their valuable gum when slits are cut in the shoots or roots. The gum is collected as flakes or ribbons. It is an emulsifier in processed foods, from mayonnaise to milkshakes.

Carob

Carob fruits
Carob *Ceratonia siliqua*
Donated by Royal Pharmaceutical Society 1983
EBC 58276
Carob trees provide a chocolate substitute from their fruit, and locust bean gum from their seeds. The Ancient Egyptians used the gum as an adhesive for mummy wrappings!

Yeheb

Yeheb nuts
Yeheb *Cordeauxia edulis*
Donated by Royal Pharmaceutical Society 1983
EBC 58258 Somalia
The yeheb tree, a native of the arid areas of Somalia and Ethiopia, produces tasty nutritious nuts. It is being tried out as a dryland crop, though its seeds contain chemicals that can cause stomach upsets.

Fenugreek

Both the seeds and leaves of fenugreek are used to flavour Indian dishes. The seeds are a characteristic ingredient of curry powders and mango chutney, and sprouted seeds give a delicate flavour to salads.

Fenugreek seeds
Fenugreek *Trigonella foenum-graecum*
Donated by Royal Pharmaceutical Society 1983
EBC 58205

Fenugreek leaves
Fenugreek *Trigonella foenum-graecum*
EBC 61396 Pakistan

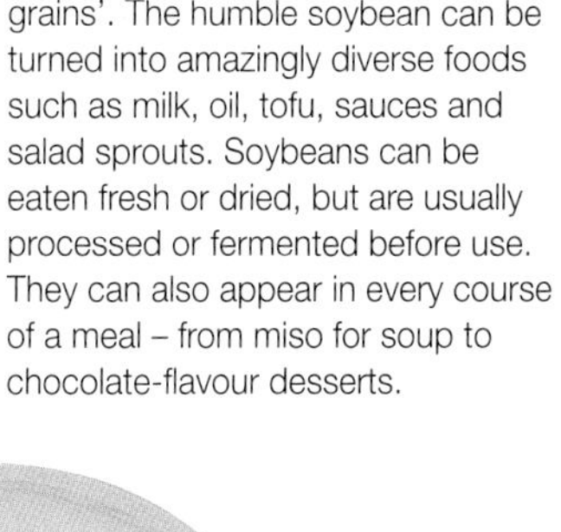

'Poor man's meat'

In China, protein-rich soybeans are often substituted for expensive meat – they are even called 'poor man's meat'. The crop has been cultivated there for over 5000 years, and is considered one of the five 'blessed grains'. The humble soybean can be turned into amazingly diverse foods such as milk, oil, tofu, sauces and salad sprouts. Soybeans can be eaten fresh or dried, but are usually processed or fermented before use. They can also appear in every course of a meal – from miso for soup to chocolate-flavour desserts.

3

Varieties of soybean
Soybean *Glycine max*
Donated by F N Howes
EBC 60462 Great Britain
Soybeans are probably native to north-eastern China, but are now grown in most warm countries, the major producer being the USA. They are one of the world's most important sources of oil and protein.

Indonesian soy sauce
Soybean *Glycine max*
1997
Soy sauce is made by fermenting a mixture of cooked soybeans and wheat, first in air and then for up to eight months in salt water. It adds a salty savoury flavour to many Oriental dishes.

Dried bean curd
Soybean *Glycine max*
1997
Bean curd is made from soy milk treated with a salt so that the protein coagulates. It can be eaten fresh or preserved by drying or freezing. The Japanese call bean curd 'tofu' and add it to soups or cook it with meat and vegetables.

Soya meal
Soybean *Glycine max*
Donated by RHM Ingredient Supplies Ltd 1980
EBC 60460
This soya meal is added to processed foods as an emulsifier that prevents the separation of fats and water. It contains a substance called lecithin which features in many margarines, chocolates and dairy desserts as E322.

Crude soya oil
Soybean *Glycine max*
Donated by British Empire Exhibition 1924
EBC 60435
Oil extracted from soybean is the major culinary oil in the United States and Western Europe. It is used as a salad and cooking oil and appears in margarines, mayonnaise and many processed foods.

Textured soya protein chunks
Soybean *Glycine max*
Donated by RHM Ingredient Supplies Ltd 1980
EBC 60445
By extruding, spinning or steam-texturising, protein extracted from soybeans can be made into products which resemble meat in texture and shape. Such textured vegetable protein (TVP) is often added as an extender to processed meats.

Soybean halves
Soybean *Glycine max*
Donated by R Sarant and Co. 1982
EBC 60454
Soybeans are split to separate the nutrient-rich seed leaves (cotyledons), which comprise the bulk of the seed, from the seed coat and embryo. The cotyledons contain 40 per cent protein and 20 per cent oil.

Bredsoy
Soybean *Glycine max*
Donated by British Soya Products Ltd 1981
EBC 60440
Soybean flour can be added to wheatflour to increase the protein content of bread. It contains various essential amino acids needed for a balanced diet. Cereals enriched with soy protein are often given to treat protein malnutrition.

10 Eating for energy

Staple crops provide most of the carbohydrate – energy food – that we need in our diets. In temperate countries, people rely on cereals such as wheat and barley, together with potatoes, for the bulk of their carbohydrate. In tropical areas, there is a far greater variety of staples: cereals such as rice and maize, starchy fruits such as breadfruit, and starchy tubers such as cassava.

Breadfruit.

Breadfruit

Breadfruit model
Breadfruit *Artocarpus* sp.
EBC 42806
The huge breadfruits, weighing up to 1.8 kg, consist of a warty rind surrounding a starchy pale yellow flesh. They rarely contain any seeds as the fruit develops without pollination taking place.

Pestle for pounding breadfruit
EBC 42860 Washington Island
In parts of Polynesia, people pound breadfruit into a pulp called 'poi', which is fermented in large pits. Poi can be kept for several years. Usually breadfruit is eaten steamed or roasted.

Sliced dried breadfruit
Breadfruit *Artocarpus altilis*
Donated by E Chitty
EBC 42762 Jamaica
Breadfruits were carried to the West Indies from their native Polynesia in the eighteenth century on a ship captained by Captain Bligh. His first attempt had ended with the infamous 'Mutiny on the Bounty'.

Potatoes

Models of potatoes – variety King Edward
Potato *Solanum* sp.
EBC 48554
Potatoes are the world's most important plant food apart from cereals. These underground stem tubers contain good quality protein, high levels of vitamin C and very little fat (until they are made into chips!).

Chuno de papa Imilla (1)
Potato *Solanum tuberosum*
Donated by S Renvoize 1979
EBC 48482 Bolivia
In the Andes, where potatoes originated, people make 'chuno' – freeze-dried potato – which can be stored for long periods. Potatoes frozen outside overnight dry quickly in the bright sunlight without thawing.

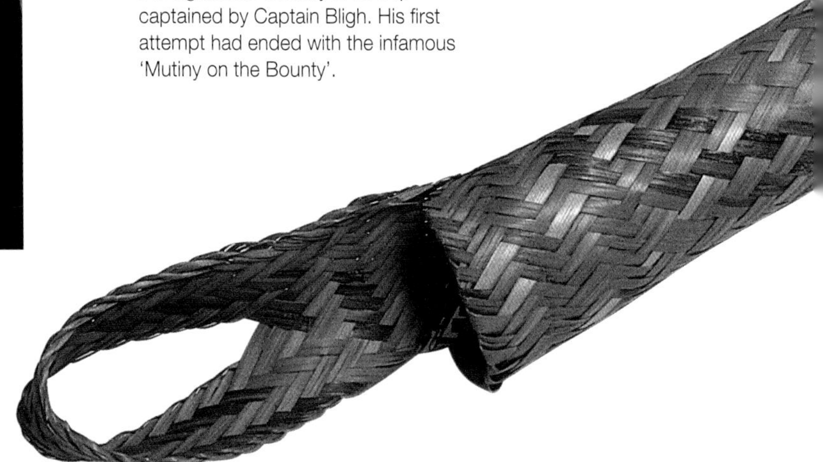

1

2

Cassava

Ho tien food paste
Cassava *Manihot esculenta*
Donated by Franco-British Exhibition 1908
EBC 44341 Indo-China
Cassava is the main food source for over 500 million people around the world – mainly in South America, where it originates, but also in Africa and Asia. It appears in many different forms.

Cassava strainer (2)
Ischnosiphon arouma
Donated by Miss Bell 1905
EBC 29514 Guyana
Careful preparation of cassava is necessary because it can produce cyanide toxins. Grated pulp is drained into a strainer and washed repeatedly. This strainer is woven from the leaf stalks of a rainforest plant.

Pearl tapioca
Cassava *Manihot esculenta*
Donated by India Museum 1880
EBC 44312 Singapore
Pearl tapioca is made from the starchy juice of cassava roots. The juice is heated on a hot metal plate until it coagulates in granules. A pudding made with tapioca and coconut milk is popular in Thailand.

Cassava grater
Cedrela odorata
Collected by W Milliken 1994
EBC 73334 Brazil
This grater would be used to shred cassava roots. The wood is dyed with annatto and the teeth are made from pieces of tin can glued with the latex of *Couma macrocarpa*.

Farinha de Mandiocca
Cassava *Manihot esculenta*
Donated by J Wetherell 1865
EBC 44346 Brazil
Grated cassava is dried and powdered before being toasted to make farinha. People sprinkle this over stews or use it to make cakes and bread.

Cassava bread
Cassava *Manihot esculenta*
Donated by A F Ridgeway 1851
EBC 44340
Griddle bread is made by pouring cassava pulp mixed with water onto a flat heated plate. The bread can be stored for long periods.

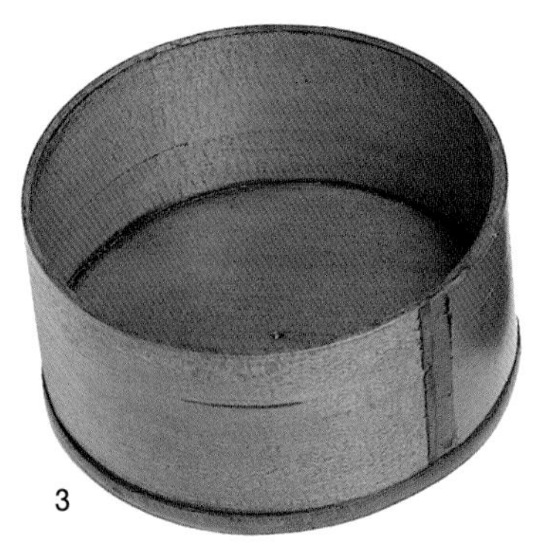

3

Rice

The custom of throwing rice at weddings reflects the oriental view of rice as sacred and a symbol of fertility. For over one and a half billion people, it is the staple food, and in many eastern cultures a meal would be considered incomplete without it. In parts of the tropics, farmers can grow two rice crops each year, supporting huge populations of people. Rice is unusual because much of it is grown in flooded fields.

Rice sticks
Rice *Oryza sativa*
1997 Thailand

Fruiting head
Rice *Oryza sativa*
Donated by India Museum 1880
EBC 32313 India

Brown rice
Rice *Oryza sativa*
1997
After harvesting, rice is threshed and winnowed to separate the grain from the husk producing nutrient-rich brown rice. Milling and polishing brown rice gives white rice, which is mainly pure starch.

White basmati rice
Rice *Oryza sativa*
1997

Short-grain rice
Rice *Oryza sativa*
1997 Italy
Unlike the long-grain rices which will only grow in tropical climates, varieties of short-grain rice can be cultivated in temperate countries such as Italy and Japan. These are used in puddings, risottos and sushi.

Noodles
Rice *Oryza sativa*
1994
EBC 73005 Sri Lanka
These noodles, or string hoppers as they are known, are made from 'instant' rice flour and cook very quickly.

Lacquer-ware rice box (3)
Rhus verniciflua
Donated by J J Quin 1882
EBC 67834 Japan
This box, made of commonest red Shunkei ware, was manufactured in the Province of Shinano.

Lacquer-ware bowl for soup or rice (4)
Rhus verniciflua
Donated by J J Quin 1882
EBC 67814 Japan

Lacquer-ware rice bowl (5)
Rhus verniciflua
Donated by J J Quin 1882
EBC 67817 Japan

Fanner of split bamboo and paper
Paper mulberry
Broussonetia papyrifera
Donated by J H Veitch 1893
EBC 42912 Japan
Farmers use this fanner to winnow cereals. They toss the mixture of threshed grains and husks up into the air so that the lighter husks blow away and the heavier grains are retained.

Wild rice grains
Wild rice *Zizania aquatica*
Donated by J Halkett
EBC 33435
'Wild rice' sold in supermarkets comes from a species of aquatic grass native to North America rather than a true rice species (*Oryza*). Wild rice was an important food source for some native North American tribes.

Harvesting rice in Madagascar.

6

Corn

Maize, or corn, has long been a vital food for people in South and Central America. When Europeans reached the Americas over 300 major varieties were already being cultivated – today, there are many thousands **(6)**.

Maize cobs
Maize *Zea mays*
Donated by E F im Thurn
EBC 33544 Guyana

Maize cobs
Maize *Zea mays*
Donated by P Cribb 1996
EBC 60674 Peru

Wafer bread
Maize *Zea mays*
Donated by National Museum of USA 1881
EBC 33897 USA
This wafer bread was made from maize by people of the Moqui Pueblos in Arizona. Other types of bread made from maize flour include corn bread and tortillas.

Popcorn ball
Maize *Zea mays*
EBC 40497
When popcorn grains are heated, they explode, turning themselves inside out. The centre of each grain contains a soft starch that produces steam when it is heated. This steam causes the grain to pop.

Mealie flour
Maize *Zea mays*
Donated by South African Produce Exhibition 1907
EBC 33708 South Africa
Maize is an important crop in South Africa, grown under the name 'mealies'. This name probably derives from the Portuguese word, 'milho', meaning grain. Mealie flour is cooked with water to make a thick porridge.

Corn meal – fine hominy
Maize *Zea mays*
Donated by J E Archer
EBC 33706 USA
Maize meal, ground to different degrees of fineness, is used to make a wide variety of different foods. Coarsely ground maize is called 'hominy' or 'grits' in America. The Italian polenta is more finely ground.

5

11 Baubles, bangles and beads

Kew botanist with seeds ready for long-term storage.

Throughout time, people of all cultures have adorned themselves with beautiful plant products, from simple seed necklaces to ornate bracelets. To many people, these plant-based decorations are as precious as gems. Their value may go beyond mere ornament, displaying the wearer's rank or having religious meaning.

Sculptural seeds

Many seeds are inherently beautiful, often finely shaped or delicately shaded. People use them as jewellery, sometimes doing little more than stringing them into anklets or necklaces. Other seeds are painstakingly carved with intricate patterns. Kew's Millennium Seed Bank, based at Wakehurst Place, aims to store samples of ten per cent of the plant kingdom by 2010.

Zulu necklace
Stangeria eriopus
Donated by K Saunders 1881
EBC 29046 South Africa
These seeds are produced by a South African cycad. Many cycads are popular ornamental plants, and are becoming endangered in the wild as they are collected for horticulture.

Necklace (1) and hat pins
Lucky beans *Afzelia quanzensis*
Donated by A King 1910
EBC 59215
These black and red seeds are often used in South Africa to make jewellery and other ornaments. The seeds of similar species are used as charms, for good luck and to ward off evil.

Seed worn by native chief
Ivory nut palm *Phytelephas macrocarpa*
Donated by J Hutchinson 1931
EBC 36113
The hard seeds of this South American palm are called 'vegetable ivory'. This natural material, which can be carved or polished, has been largely replaced by plastics, but there is now renewed interest in it.

Yanomami necklace
Renealmia alpinia
Donated by W Milliken 1994
EBC 73336 Brazil
To make this necklace, the seeds were strung onto a cotton thread whilst they were still soft. The necklace was made in the village of Watoriketheri, located in Roraima State, Brazil.

Necklace (2)
Velvet plant *Lacunaria jenmanii*
Donated by E F im Thurn 1889
EBC 66694 Guyana
This South American plant gets its name from the covering of fine soft hairs on the seeds.

Decoratively strung seeds
Lead tree *Leucaena leucocephala*
EBC 59048
The lead tree's shiny brown seeds are stuck together to make beads. The tree originates in Mexico, where people also use its wood for fuel and paper pulp and its leaves as animal forage.

Anklet
Autranella congolensis
Collected by N Rumball 1995
EBC 73574 Cameroon
Villagers in Cameroon adorn themselves with these decorative anklets during tribal dances.

3, 4, 5, 6, 1, & 2 (left to right; see overleaf)

ob's tears

ob's tears, or pit-pit beads, are the
alse fruits of a tropical grass. With
neir hard shiny shells, in shades of
ream or grey, they are ideal for
naking jewellery. Other varieties of
ob's tears are eaten as a cereal.

ecklaces
ob's tears *Coix lacryma-jobi*
onated by India Office 1898
BC 31800 India

ecklace including sea heart
ob's tears *Coix lacryma-jobi*
ea heart *Entada* sp.
BC 40523 Kenya

rmband
ob's tears *Coix* sp.
881
BC 31803 New Guinea

Necklaces
Castor oil plant *Ricinus communis*
Job's tears *Coix lacryma-jobi*
Donated by E Cook 1985
EBC 40136
The darkly mottled seeds of the castor oil plant complement the ivory-coloured Job's tears. Castor oil seeds contain the poison, ricin. Doctors are investigating ways of using it to treat cancer.

Dangerous jewellery

Bracelet of seeds
Abrus precatorius
Collected by D Goyder 1996
EBC 73636 Kenya
These vivid red seeds, commonly known as crab's eyes or jequerity seeds, make colourful beads. However they are extremely poisonous! Kew answers many enquiries about jewellery made from them.

Seed and silver earrings
Abrus precatorius
1997
EBC 73824 Peru

Hat pins
Water chestnut *Trapa natans*
Crab's eyes *Abrus precatorius*
Donated by A King
EBC 54895
The starchy kernels in the triangular water chestnut fruits are more commonly used as food. In India they are eaten boiled and roasted or made into flour, while in China they are a popular vegetable.

7

Ornaments

Bangles of lac
Ficus sp.
Donated by Paris Exhibition 1900
EBC 43187 India
Lac comes from the resinous coating of various insects. In India, these insects are nurtured on fig trees (*Ficus* spp.). Most lac is used in varnishes, but some goes to make jewellery.

4

Beads made from lac
Ficus sp.
Donated by India Museum 1880
EBC 43201 India

Rings made from fruits
Java almond *Canarium* sp.
Donated by F N Howes 1928
EBC 63221 Indonesia
These rings are carved from the woody layer that surrounds the seeds in Java almond fruits. A similar hard layer occurs in peach fruits, encasing the kernel.

Leg ornament
Snuff box tree *Oncoba spinosa*
Donated by A W Kranks 1892
EBC 66978 Zimbabwe
The fruits making up this leg ornament rattle together as the wearer moves, so they make music as they dance.

Bracelet
Peach *Prunus persica*
Donated by W R Carles 1898
EBC 57363 China
Peach stones have a hard stony layer around the central kernel. These stones have been carved into figures which represent the disciples of Buddha.

Necklace (3)
Blue fig *Elaeocarpus angustifolius*
Donated by L Bernays 1849
EBC 64704 Australia
When the spherical blue fruits of the blue fig fall to the rainforest floor, the thin outer layer of flesh rots away to leave these beautifully sculpted stones which surround the seeds.

Necklace of Irish bog oak (4)
Oak *Quercus* sp.
Donated by Lady Doneraile 1857
EBC 37986 Ireland
Occasionally, oak logs fall into bogs or lakes and, instead of decaying, are preserved by airless conditions in the waterlogged soil. The blackened wood that results is much prized for making jewellery and furniture.

Necklace made from bark (5)
Birch *Betula* sp.
Donated by Norwegian Forestry Institute 1981
EBC 42472 Norway
Birch trees grow in northern regions of the world. Many have flexible, waterproof bark which peels away from the trunk. This bark has been used for different products from paper to canoes, as well as jewellery.

Necklace of sarpat straw (6)
Sarpat *Saccharum arundinaceum*
Donated by J F Duthie 1886
EBC 32427 India
Sarpat straw is strong and elastic – villagers in India use it to make mats and ropes as well as jewellery.

Zulu necklace
Protea *Protea* sp.
Donated by K Saunders 1881
EBC 45023 South Africa
This necklace is made from protea flower receptacles – the tip of the flower stalk that supports the petals and other flower parts. Many proteas have spectacular flowering heads, often dripping with nectar.

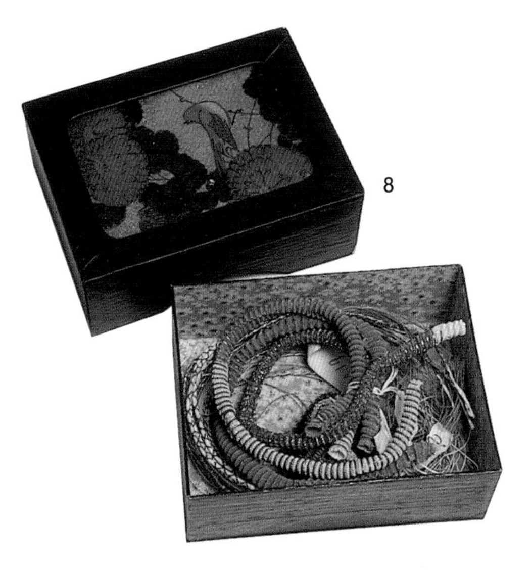

8

Comb (7)
Oenocarpus bacaba
Collected by W Milliken 1994
EBC 73333 (Brazil)
This comb was made by the Wai-Wai tribe. Its teeth are from the leaf bases of the palm *Oenocarpus bacaba*, woven with the leaves of another palm (*Mauritia flexuosa*) and dyed cotton (*Gossypium barbadense*).

Carved fruits
Chinese olive *Canarium pimela*
Donated by Foreign Office 1893
EBC 55509 China
The hard shells that surround the kernels have been elaborately carved into figures from Chinese folklore.

Woven armlets
Screwpine *Pandanus* sp.
Donated by R B Comins 1890
EBC 34811 Solomon Islands
Screwpines get their name from the distinctive spiral arrangement of their leaves. Their fibrous leaves have been an important source of weaving material on many Pacific Islands.

Paper hair decorations

Japanese women ornament their hair with paper decorations. The white inner bark of paper mulberries (*Broussonetia* spp.) is used to make many different forms of paper, both ornamental and functional.

Box of hair ornaments made of paper (8)
Paper mulberry
Broussonetia kazinoki
Collected by H Parkes 1871
EBC 42845 Japan

Threads of paper to tie hair (9)
Paper mulberry
Broussonetia papyrifera
Donated by J H Veitch 1893
EBC 42856 Japan

Aboriginal ornaments

Australian Aboriginal people wore ornaments like these gnulga and mulgeddies at corroborees and war dances.

Gnulga or hair ornaments
Gum tree *Eucalyptus* sp.
Donated by Dr Clement 1898
EBC 55335 Australia

Mulgeddie or nose ornaments
Gum tree *Eucalyptus* sp.
Donated by Dr Clement 1898
EBC 55334 Australia

9

12 Nature's bounty – coconut

Coconut palms are amazingly useful: their leaves are used for clothing, mats, baskets and roofing; their fruits provide food, drink, oil, containers, fibre for ropes and doormats; and their wood helps build houses and boats.

Common mat made of leaves
Coconut palm *Cocos nucifera*
Donated by G H K Thwaites 1874
EBC 35471 Sri Lanka

Cable made of tarred coconut coir
Coconut palm *Cocos nucifera*
Donated by India Museum 1880
EBC 35389 India

Spoons using coconut shell
Coconut palm *Cocos nucifera*
EBC 35387 Sri Lanka

'Grass' skirt made from palm leaves (1)
Coconut palm *Cocos nucifera*
Donated by F W Christian 1899
EBC 73292 Federated States of Micronesia

Fishing basket
Coconut palm *Cocos nucifera*
Donated by G H K Thwaites 1874
EBC 38611 Sri Lanka

Walking stick
Coconut palm *Cocos nucifera*
Donated by J D Hooker
EBC 37504 India

Copra for coconut oil
Coconut *Cocos nucifera*
Donated by British Empire Exhibition 1924
EBC 35440 Trinidad

Coconut fibre for brushes
Coconut palm *Cocos nucifera*
Donated by Chubb, Round and Co. 1885
EBC 3537

1

Coconut, Madagascar.

Nature's bounty – pineapple 13

Luscious pineapple fruits are not only delicious to eat but can also be used to tenderise meat, because they contain an enzyme that breaks down proteins. Pineapple leaves provide a strong, fine fibre suitable for fabrics and string.

Twine from pineapple fibre
Pineapple *Ananas comosus*
EBC 29684 India

Cloth of pineapple fibre
Pineapple *Ananas comosus*
Donated by Government of India
EBC 29684 India

Fibre stages
Pineapple *Ananas comosus*
Donated by Acting Consul Parker 1891
EBC 73300 China

Cigar case from strips of pineapple
Pineapple *Ananas comosus*
Donated by J Henshall
EBC 48317 Philippines

Meat marinade
Pineapple *Ananas comosus*
Donated by RBG Kew staff 1984
EBC 29670

Shirt from pineapple fibre (1)
Pineapple *Ananas comosus*
Donated by J Bowring 1859
EBC 73091 Philippines

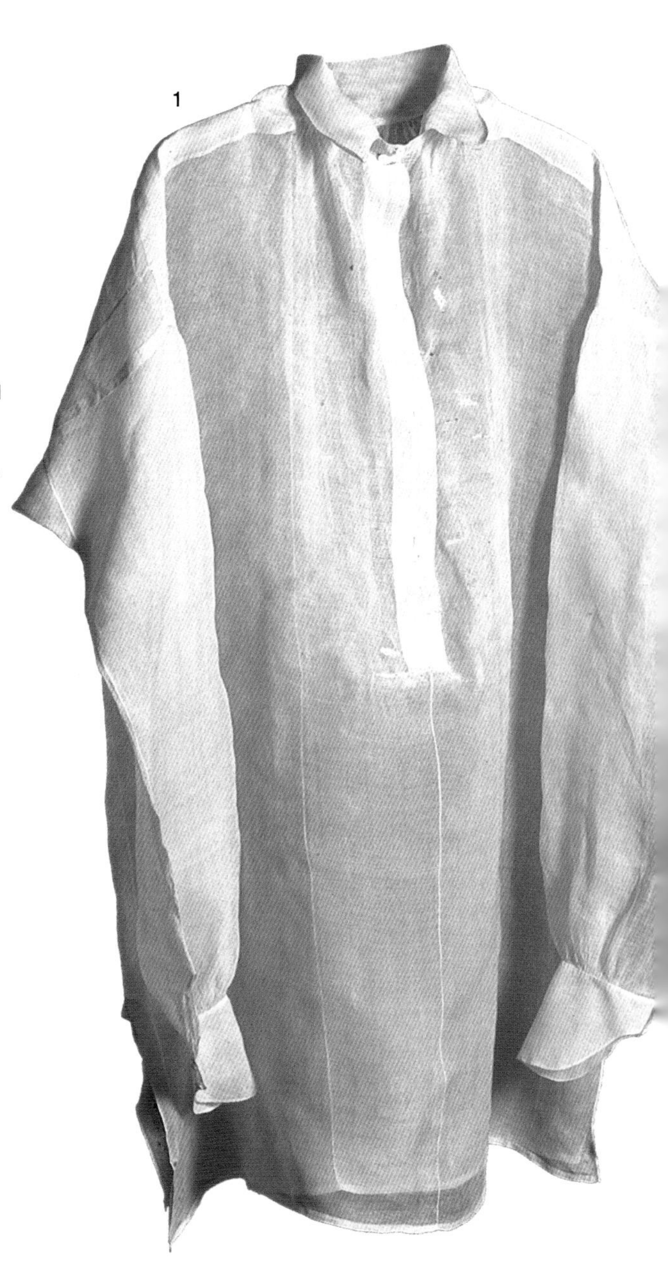
1

14 Head to toe

Weaving a hat from leaves of *Copernicia prunifera*, Ceará, Brazil.

Plants are used across the globe to create hats and shoes that are either decorative or functional – sometimes both. Remarkably similar styles have appeared in completely different places.

Hats

Hats protect us from pouring rain and burning sun, or just add a touch of style. A good sunhat provides shade while allowing air to circulate. Hats woven from the leaves or stems of local plants do the job perfectly, and come in all shapes and sizes. In Northeast Brazil, hats are woven from local palm leaves.

Yemeni hat
Date palm *Phoenix dactylifera*
Collected by M van Slageren 1997
EBC 73867 Yemen
There are over 800 different recorded uses of date palms. Their fruits provide energy-rich food and their leaves are also used to make a huge range of products from hats to roofs.

Lady's hat of maize husks
Maize *Zea mays*
Donated by J Burtt Davy 1920
EBC 33910 South Africa
In South Africa, maize is an important staple food for many people. The husks surrounding the cobs are a by-product put to good use in making hats and other basketry items.

Hat made from cycad leaves (1)
Cycas revoluta
Donated by J H Veitch 1892
EBC 29021 Japan
The design of this hat is particularly ingenious, using complete cycad leaves. The leaf stalks form the ribs of the hat, and the leaflets that they bear are interwoven to make its fabric.

Basket hat made from split twigs (2)
Willow *Salix* sp.
Donated by US National Museum 1891
EBC 41419 USA
This hat was worn by the Paiutes of southern Utah. Willow shoots are very flexible and are ideal for basketry. Usually they are stripped before use but basket-makers sometimes make a feature of the coloured bark.

Hat showing plaiting methods
Panama hat plant *Carludovica* sp.
Donated by W Fawcett 1903
EBC 73304 Jamaica
Each Panama hat is made from 10–12 fan-shaped leaves. These are gathered while young and cut into strips. As the strips dry, they curl up into cylinders that are soft and pliable enough to weave.

Hat
Coconut palm *Cocos nucifera*
Donated by D Hall 1988
EBC 38690 Ghana
Reaching up to six metres in length, coconut palm leaves bear 200–250 leaflets. These are used throughout coastal tropical regions in a wide variety of woven products, from hats to baskets.

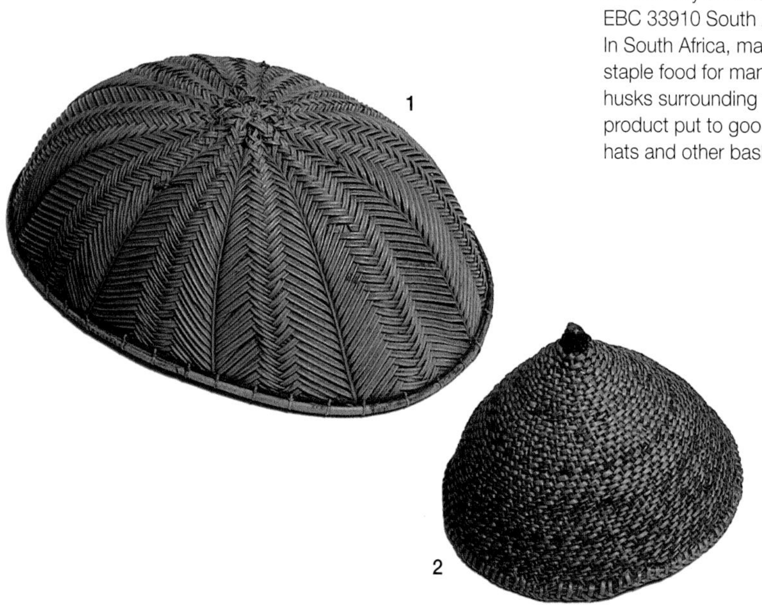

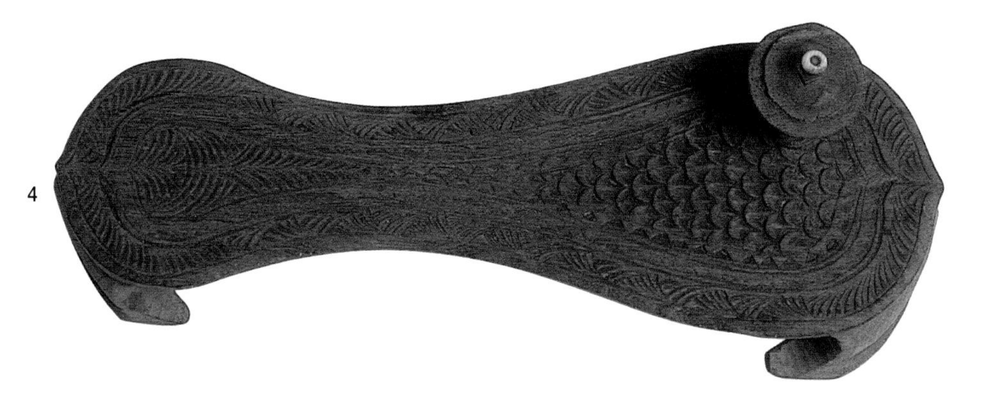

4

Sola pith hat
Sola tree *Aeschynomene aspera*
Donated by N Martland 1997
EBC 73876 India
Pith helmets are seen as the typical attire of many tropical explorers, including botanists. They were made from the wood of the sola tree, which is extremely lightweight but screens the wearer from the sun's heat.

Helmet
Rattan *Calamus* sp.
Collected by H M Burkill 1908
EBC 35286 Bangladesh
Rattans are climbing palms from the rainforest whose stems provide cane for making furniture, baskets and mats. Kew research aims to identify species suitable for cultivation.

Hat made from screwpine leaves
Screwpine *Pandanus* sp.
Donated by A E Evans 1906
EBC 34844 Ghana
Screwpines take their name from the spiral arrangement of their leaves around the stems. Dried leaves are used to make baskets and mats, while bags and ropes are manufactured from leaf fibres.

Cork hat (3)
Cork oak *Quercus* sp.
Donated by J A Henriques 1882
EBC 37827 Portugal
Lightweight cork provides insulation. It is made from the bark of the cork oak, removed from the trunk in cylinders. If it is harvested carefully, cork can be cut away every eight years.

Shoes

Although shoes are most often made from animal skins, plants also feature in footwear. Sturdy wooden clogs keep feet dry, while delicate fabric slippers are saved for special occasions.

Getta
Paulownia tomentosa
Quercus mongolica
Donated by J H Veitch 1893
EBC 41866 Japan
Two different woods have been used to construct this shoe – the flat sole is made from the ornamental tree *Paulownia tomentosa* while the uprights are of an oak, *Quercus grosseserrata*.

Wooden sandal (4)
Teak *Tectona grandis*
Donated by H Cape 1861
EBC 37960 India
This sandal was worn by Indians during cooking and bathing. Teak is a tropical timber tree with special decorative qualities, widely used for furniture and building work.

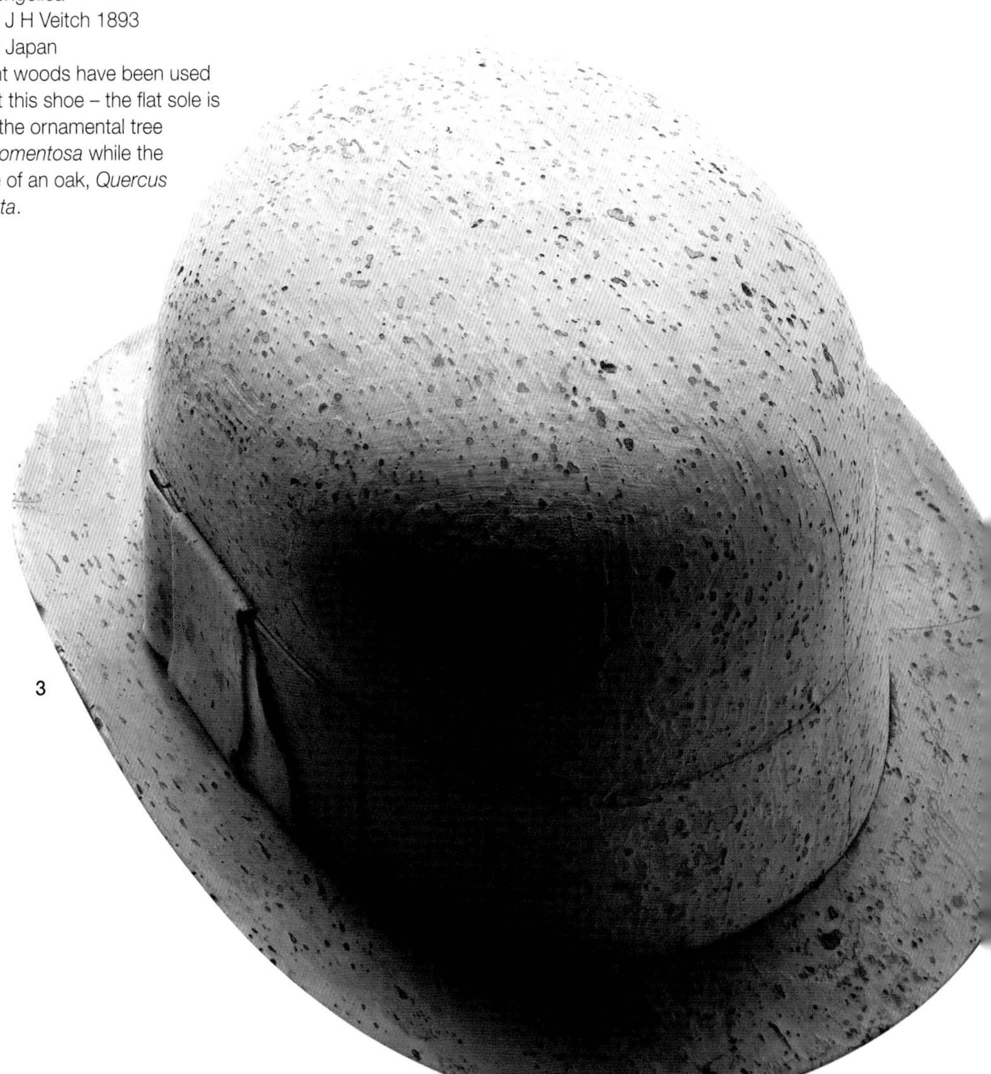

3

5

Clogs

Clogs, or sabots as they are sometimes known, carved from single blocks of wood, are among the simplest forms of footwear. Tough, light willow wood is ideal for these hard-wearing shoes.

Sabots
Willow *Salix* sp.
Donated by T Watters 1888
EBC 41412 Korea

Dutch sabots
Willow *Salix* sp.
Donated by T Clayden
EBC 41403 Netherlands

Half-made shoes of willow
Willow *Salix* sp.
Donated by Hungarian Exhibition 1908
EBC 41404 Hungary

6

Rubber shoes

Early explorers in the Amazon rainforest described a local waterproof foot covering made of a white 'sap' now known as rubber latex. Rubber still features in footwear, from tennis shoes to wellington boots.

Rubber ornamental shoe
Rubber *Hevea* sp.
Donated by Mackintosh and Co. 1853
EBC 44133

Rubber shoe
Rubber *Hevea* sp.
Donated by Mr Churchill 1898
EBC 44195 Brazil

Lace bark slippers

Lace bark
Lace bark tree *Lagetta lagetto*
Donated by G A Prinsep 1979
EBC 44963 Cuba

Slippers made of lace bark
Lace bark tree *Lagetta lagetto*
Donated by Dr Bowerbank 1827
EBC 67770 Jamaica
These slippers are made of the delicate, mesh inner bark of the lace bark tree. Their soles are formed from 'cocoa nut bark and India rubber fibre'.

7

Shoe of bark (5)
Birch *Betula alba*
Donated by Dr Anderson
EBC 42416 Sweden
Peasants living in northern Sweden wore shoes like this to keep their feet dry when visiting swampy meadows. The waterproof birch bark is flexible enough to weave.

Shoes (6)
Marram grass *Ammophila arenaria*
EBC 31691 Great Britain
These shoes were made at Great Yarmouth in Norfolk where marram grass grows on coastal sand dunes. The grass is often planted deliberately to control erosion because its roots stabilise the sand.

Sandal (7)
Lime *Tilia* sp.
Donated by A Henry 1890
EBC 64851 China
Lime trees produce a fibrous inner bark suitable for coarse cordage and matting.

Tapping rubber in Cameroon.

15 Hunting, shooting and fishing

Plants themselves are a major source of human food, but people also use them to catch animals to eat. The methods devised to catch fish have been particularly ingenious, using traps and poisons as well as nets and lines. On land, bows and arrows, blowpipes, spears and guns are used to hunt prey.

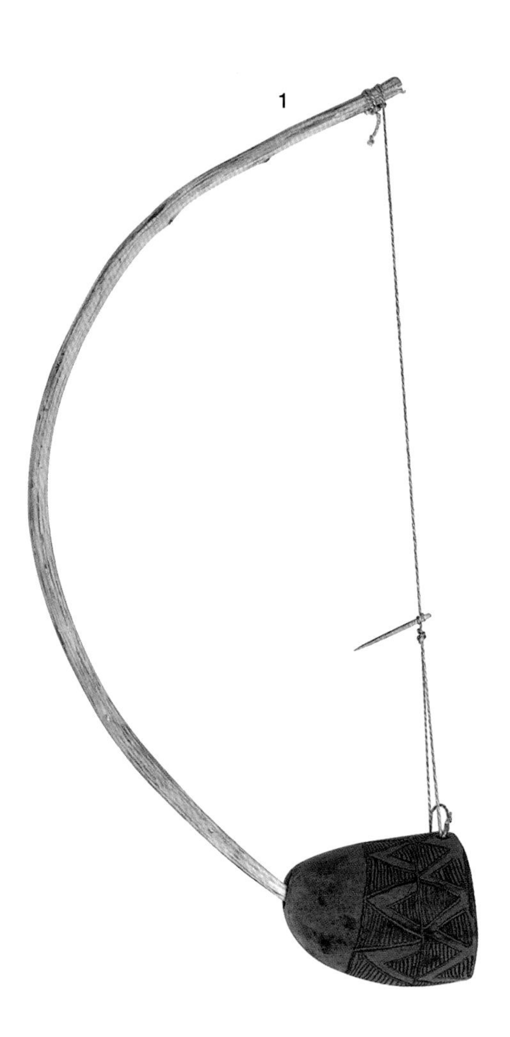

From poison to protection

In many tropical countries, people have discovered various plants that contain chemicals capable of killing fish without making them unsafe to eat. Some of these chemicals have found new uses as insecticides, protecting crops without harming the environment. Scientists at Kew are actively involved in research to identify other plant products able to control pests.

Blowpipe
Bamboo *Arundinaria* sp.
EBC 40759 Pakistan
With their hollow woody stems, bamboos are ideal materials for blow-pipes. They are used to project lightweight darts which are frequently tipped with poison to kill the prey quickly.

Quiver with arrows
Bacaba palm *Oenocarpus distichus*
EBC 38664
Bacaba palm grows in lowland parts of the Amazon rainforest. Unusually long arrows, like these, are used by the Yanomami peoples to shoot fish as well as birds and small mammals.

Quiver of poisoned blowpipe darts
Donated by L Wray 1891
EBC 42727 Malaysia
These darts are coated with ipoh, a fast-acting poison. The source of the poison is believed to have been a member of the mulberry family, the Moraceae.

Quiver with poisoned arrows
Bamboo
Donated by S W Silver 1878
EBC 40633 Polynesian Islands
Fast-acting arrow poisons made from a wide range of plants are used in many parts of the tropics. Some are valuable drugs – compounds from a South American arrow poison are used as muscle-relaxants before surgery.

Stems used as fish poison
Barbasco *Lonchocarpus* sp.
Donated by International Exhibition 1862
EBC 60631 Guyana
For centuries, various South American peoples have used the pounded stems of barbasco as a fish poison. The chemical responsible, rotenone, has since been found to be an insecticide.

Tuba root used as fish poison
Tuba *Derris elliptica*
Donated by Singapore Botanic Garden 1922
EBC 60288 Singapore
Tuba root has long been used in South East Asia against hair lice and as a fish poison. By 1877, it was being cultivated as a source of insecticide for pest infestations of crops.

Derris powder
Tuba *Derris* sp.
1997
The insecticidal compound extracted from tuba roots is called rotenone. Unlike many synthetic pesticides, it is biodegradable and does not persist in the environment.

3

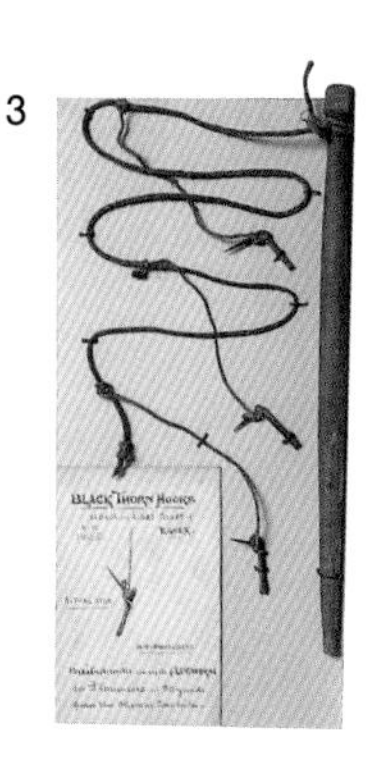

4

Mousetrap (1)
Shaft: *Grewia plagiophylla*
Twine: *Agave sisalana*
Cup: baobab *Adansonia digitata*
Trap spine: *Premna hildebrandtii*
Bait: maize *Zea mays*
Collected by J Morley and S Davis 1995
EBC 73480 Kenya
This mousetrap is used in parts of Kenya to catch small mammals for food.

Bamboo bow strung with *Bauhinia* bark
Bauhinia vahlii
EBC 61569
Bauhinia vahlii is an enormous climber which grows widely in India. Its fibrous inner bark is used to make rope and string.

Crossbow with string of plant fibre
Indian laburnum *Cassia fistula*
EBC 37972 India
Wood for crossbows must be flexible but tough. Indian laburnum wood, used for this crossbow, is strong and tough, although it tends to crack during seasoning.

African spears
Oxytenanthera abyssinica
Donated by Mr Franks 1881
EBC 38661 Nigeria
This bamboo, widespread in dry forests in tropical Africa, produces woody stems up to 15 m tall. Thinner stems are valued for bows, arrows and spears, while thicker stems are used for huts and furniture.

Boomerang
Acacia sp.
EBC 59890 Australia
Boomerangs, made from curved hard pieces of wood, were used as weapons by Aboriginal peoples of Australia. When thrown, the weapon followed a curved path, and could be made to return to its owner.

Halibut hook (2)
Western hemlock *Tsuga heterophylla* 1904
EBC 28575 Canada
The dried wood of the western hemlock is extremely hard. It forms a sturdy base for this fish hook made by the Salish people of Vancouver Island. They also ate the tree's inner bark, fresh in spring or dried in winter.

Blackthorn hooks (3)
Blackthorn *Prunus spinosa*
Donated by R T Pritchett 1896
EBC 57362 Great Britain
The sharp spines of the blackthorn were used to fish for flounders on the Essex coast. Each spine was baited with lugworm and the line of hooks was laid out about 50 m from shore.

Fishing rope
Sunn hemp *Crotalaria juncea*
Donated by G H K Thwaites 1874
EBC 60192 Sri Lanka
Sunn hemp is strong and durable and is widely used to make fishing lines and nets. During World War 2, it was used for camouflage nets.

Eelpot (4)
Willow *Salix* sp.
EBC 41411 Great Britain
Eelpots made of interwoven willow rods were once common along the River Severn and in the Fenlands of East Anglia.

Hand-fishing net
Gnetum scandens
Donated by P Vaux 1902
EBC 27110 India
The flexible stems of the large evergreen climber *Gnetum scandens* provided the fibre for this net made by the women of the Andaman Islands. They used the nets to catch fish and prawns.

Telescopic fishing rod
Japanese female tonkin *Pseudosasa japonica*
Donated by H N Moseley 1879
EBC 34073 Japan
Several species of bamboo have been used to make fishing rods. *Pseudosasa japonica* made low quality rods, but higher quality rods for fly-casting consisted of strips of tonkin cane (*Arundinaria amabilis*).

Powder flask
Bamboo
EBC 40632 Africa
Powder flasks held the gunpowder that was needed to fire early guns. The gunpowder itself often contained charcoal made from wood.

Shotgun stock of walnut heartwood
Walnut *Juglans regia* 1893
EBC 37713
Finely grained walnut wood is the preferred timber for making the butts and stocks of high-quality hunting rifles and guns, because it does not shatter under the recoil shock. Furniture makers also value the wood.

Shotgun stock
Walnut *Juglans regia*
EBC 37723

2

16 To have and to hold

Containers have many uses – they can be used to store grains and other foods after harvest, to serve up meals and drinks, or to carry clothes or produce. People have found an enormous number of plants that provide materials suitable for containers, which range in style from simple baskets to immensely ornate lacquer boxes.

1

Versatile bamboo

Bamboos in baskets and boxes

Skilled craftspeople find many uses for bamboos. These giant grasses, which grow in many parts of Asia, Africa and the Americas, have hollow woody stems with cross-walls at intervals. A segment, sealed at one end with a cross-wall, is a natural storage vessel, while flexible strips of split bamboo can be woven into all manner of baskets. Giant bamboos can grow over 25 metres high! In China, livestock are transported to market in woven bamboo boxes. Strips of bamboo are used in Cambodia to form delicate rice baskets.

Two bamboo measures
Bamboo *Bambusa bambos*
Donated by India Museum 1880
EBC 40627

Basket
Bamboo *Bambusa bambos*
Donated by Indian Forest Department 1863
EBC 33961 India
The woody stems of *Bambusa bambos* are used to make different types of container. Segments of the stem are used to hold liquids and grains, while split stems are woven into multipurpose baskets.

Cigar case of Kayon chief
Bamboo
Donated by India Museum 1880
EBC 48272 Malaysia

Basket made of bamboo and rattan
Weave: *Neomicrocalamus andropogonifolius*
Frame: *Calamus acanthospathus*
Collected by S Barrow 1997
EBC 73828 Bhutan
This basket was purchased by a Kew botanist during an expedition to Bhutan to survey rattans (climbing palms) and bamboos. The weave is made of bamboo and the frame of rattan.

Beer container
Internal cylinder: *Dendrocalamus hamiltonii*
Outer weave: *Neomicrocalamus andropogonifolius*
Stitching: *Calamus acanthospathus*
Collected by R Pradhan, S Barrow and C Stapleton 1997
EBC 73827 Bhutan
Bamboos feature in both the inner cylinder of the beer container, made of a section of the woody stem, and the outer weave, made from strips of a different type of bamboo.

2

Bowls made from calabashes (1)
Calabash *Crescentia cujete*
Donated by R Temple
EBC 46482 Honduras

Packee water bottle
Calabash *Crescentia cujete*
Donated by Miss Courtney Bell 1905
EBC 46499 Jamaica
Calabashes are made from the dried fruits of a tropical American tree *Crescentia cujete*. Some fruits can hold up to 4 litres of liquid. Many calabashes are highly decorated with dyeing or engraving.

Gourd and cover
Pumpkin *Cucurbita pepo*
Donated by J Wetherell
EBC 54819 Brazil
Gourds are fruits of various members of the cucumber family, the Cucurbitaceae. After drying and removal of their seeds, they are used as containers – for carrying and storage – in many parts of the tropics.

Lacquer food box (2)
Rhus verniciflua
Donated by J J Quin 1882
EBC 67826 Japan
Lacquer is made from the resin of various species of *Rhus* (or *Toxicodendron*). The trees are native to China, but they, and the art of lacquerware, are also found in Japan. Making lacquerware is a very complicated process.

Panela pot and lid (3)
Pottery tree *Licania octandra*
Collected by R Spruce 1849
EBC 37796 Brazil
Pots made from the pottery tree's powdered bark mixed with clay can withstand great heat. This pot was collected in the mid-nineteenth century; similar pots are still made today.

Swill or osier basket
Willow *Salix* sp.
EBC 40663 Great Britain
This basket was used to carry herrings at Great Yarmouth. It is made from osiers, the long straight shoots that develop when willows are cut back to ground level each year – a process known as coppicing.

Willow basket
Willow *Salix* sp.
Donated by J Burtt Davy 1897
EBC 40660 USA
The Pima people of Arizona made baskets like this from foundation coils wrapped and sewn together with willow bark. They can hold water because the bark swells up when wet, sealing any gaps.

Siberian grain basket
Birch *Betula* sp.
Donated by Dr Lockhart 1874
EBC 42434 Russia
Birch bark is not only flexible but also waterproof. Many people in northern temperate countries, where birches grow, have used the bark to fashion diverse storage vessels.

Portmanteau
Paper birch *Betula papyrifera*
Donated by Professor Saunders 1888
EBC 67803 Canada

3

4

Basket made of finely split leaf stalks
Double coconut *Lodoicea maldivica*
Donated by Mrs Morris 1879
EBC 35810 Seychelles
Double coconuts, or coco-de-mer, are only found wild in the Seychelles. Their leaves are used to make baskets and hats, but the trees are best known for their huge seeds – the largest in the plant kingdom.

Lady's bag woven from aloe fibre
EBC 36717 France
This delicately woven bag is made from a fibre that the makers called aloe fibre. It is more likely to be either Mauritius hemp (*Furcraea gigantea*) or from a species of agave.

Pannier for carrying produce
Ischnosiphon obliquus
Collected by G T Prance 1987
EBC 73506 Brazil
In parts of Brazil, people use panniers like this one to carry produce, such as crude rubber, from the forest and their fields back to their villages.

Wooden salt cellar (4)
Black poplar *Populus nigra*
Donated by Hungary Exhibition 1908
EBC 41351 Hungary
This salt cellar was made by hollowing out a length of trunk. Wooden containers are often formed as boxes from several flat pieces of wood.

Basket made of split wood
American ash *Fraxinus americana*
Donated by Professor Saunders 1888
EBC 50360 North America
Thin strips of wood, carefully split off a piece of timber, have been interwoven to produce this decorative basket.

Game bag made from roots
White mulberry *Morus alba*
Donated by A Henry 1895
EBC 43336 Taiwan
Mulberry trees are the source of several different fibres. Silkworms feed on their leaves, a silky white fibre can be extracted from their bark, and their twigs and roots are woven into baskets and bags.

Words and pictures 17

All written or printed communication still relies on plants, even in today's computer age – when print-outs appear on plant-based paper. Charcoal, made from partially burnt wood, was one of the earliest tools for drawing and is still used today. Other plant products appear in pens, ink and pencils, and carved or engraved wooden blocks are still used for printing some illustrations.

Pulp fiction

The word 'paper' comes from the word papyrus, the Ancient Egyptian writing material made from strips of the papyrus stem. Tree bark and leaves have also been used as paper but today nearly all paper is made from woodpulp. Bulk paper-making is a huge industry, which uses up forests of trees every year. Recycling paper helps to conserve some of these trees. To produce paper, wood pulp is mixed with water to make a slurry; as the water drains away, the wood fibres settle into a sheet.

1

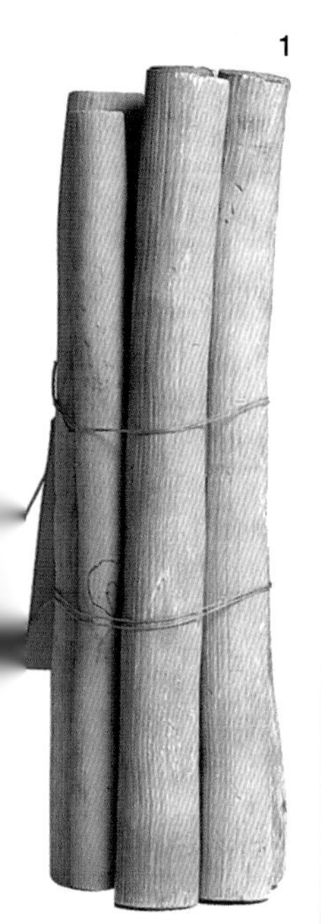

2

Paper from bananas

Musa textilis is closely related to the banana plant. Fibre for paper-making is extracted from its leaf sheaths. As well as fine handmade papers, these fibres are used in brown Manila envelopes and tea bags.

Paper stock
Manila hemp *Musa textilis*
Donated by T Routledge 1875
EBC 29617

Paper
Manila hemp *Musa textilis*
Donated by E Koretsky 1985
EBC 29618 USA

Rice-paper

Despite its name, rice paper is not made from rice, but from spirally-cut strips of the stem pith of *Tetrapanax papyrifer*, a plant related to ivy. In Asian countries, it is used for fine artwork.

Pith of rice-paper plant (1)
Rice-paper plant *Tetrapanax papyrifer*
Donated by Japan British Exhibition 1910
EBC 40128 Taiwan

Paintings of rice-paper production – painted on rice-paper (2)
Rice-paper plant *Tetrapanax papyrifer*
Donated by T Watters 1882
EBC 33725 Taiwan

3

The Polynesian Gazette.

LEVUKA, TUESDAY, OCTOBER 27, 1885.

Papyrus
Papyrus *Cyperus papyrus*
Donated by Mrs Brightwen 1891
EBC 73317
Cyperus papyrus is a sedge that grows in wet areas in tropical Africa. Strips of the white pith from its stalks have been made into writing materials for more than 5000 years.

Singhalese book of talipot palm leaves
Talipot palm *Corypha umbraculifera*
Donated by G W Cripps
EBC 35333 Sri Lanka
Palm leaves have long been a writing material. The talipot palm's long broad leaflets are cut and boiled in milk or water before being polished. Individual sheets strung onto a cotton cord form a book.

Boxwood

Box is the preferred wood for the production of wood engravings because it is dense, even-textured, and can be very finely worked. The 'negative' illustration is engraved across the grain of the wood.

Wood prepared for engraving
Box *Buxus sempervirens*
Donated by R J Scott
EBC 43794

Wooden block engraved with figure of man
Box *Buxus* sp.
Donated by W G Smith 1873
EBC 37751

Paper mulberry

Asian papermakers produce a diverse range of papers from the bark of the paper mulberry – from oiled paper for waterproof clothing to delicate, translucent washi papers suitable for calligraphy.

Coarse oiled paper
Paper mulberry *Broussonetia papyrifera*
Donated by T Watters 1888
EBC 42939 Korea

Newspaper – *The Polynesian Gazette* (3)
Paper mulberry *Broussonetia papyrifera*
1855
EBC 42906 Polynesia

Washi – Japanese handmade papers
Paper mulberry *Broussonetia papyrifera*
Donated by Morita Japanese Paper Co. 1983
EBC 42940 Japan

Chinese printing block (4)
Red pear *Pyrus betulifolia*
Donated by W R Carless 1897
EBC 57406 China
Wooden printing blocks have been used for at least 2000 years. Hard and even-grained pear wood is ideal for intricate carving, as on this block where the complex Chinese characters stand out for printing.

Engraving blocks
American holly *Ilex opaca*
Donated by R J Scott 1884
EBC 62866
Holly is a hard wood that can be used for very detailed carving. On this block the image has been cut across the grain of the wood. In technical terms, this produces a wood engraving.

Paper
Edgeworthia gardneri
Donated by R Alcock
EBC 44924 Japan
Although it is soft and supple, handmade mitsumata paper is strong enough to be used in Japanese banknotes. It is made from the white inner bark of *Edgeworthia gardneri*.

Inner bark
Edgeworthia gardneri
Donated by Japan-British Exhibition 1910
EBC 44928 Japan

Booklets printed on bark
Paper birch *Betula papyrifera*
Donated by Mrs James 1883
EBC 48340 USA
Paper birch takes its name from the papery white outer layers of bark that peel away in strips. These booklets are printed on this flexible bark.

Sections of wood and engraved cards
Sugar maple *Acer saccharum*
1895
EBC 62248
Small sheets of wood are occasionally used for printing. Usually, however, wood is made into pulp for paper production.

Artist's charcoal
Willow *Salix triandra*
Collected by H D V Prendergast 1996
EBC 73778 Great Britain
In late November, straight young willow shoots are harvested and packed into iron firing boxes, which are then sealed and heated. Under these airless conditions, the willow rods are slowly converted to charcoal.

Pencils
Cedar
EBC 28814, 28815
Pencils are made from graphite enclosed in cedar wood. Originally this was from the pencil cedar (*Juniperus virginiana*), but stocks declined due to overharvesting.

Ink made from marble galls
Oak *Quercus robur*
Donated by J Stubbs 1975
EBC 41857
Marble galls are produced on oak trees by parasitic insects. To make ink, crushed galls are mixed with boiling water and strained after a day. The resulting liquid is mixed with an iron salt and gum arabic (*Acacia senegal*).

4

Marking nut
Marking nut tree *Semecarpus anacardium*
Donated by Paris Exhibition 1900
EBC 62150 India
These fruits yield an oily black juice that, when mixed with lime, forms an indelible ink. It has traditionally been used for marking linen and, in India, was also used for tattooing.

Writing desk made of birch bark
Birch *Betula alba*
Donated by Paris Exhibition 1855
EBC 42411 Sweden

Papier-mâché and lac pen box
Donated by India Museum 1880
EBC 68579 Pakistan
This pen box was made from paper, in the form of papier-mâché, and coated with lac, produced by an insect cultured on various species of *Ficus* and *Ziziphus*.

Writing box in 'Chamoku' style
Rhus verniciflua
Donated by India Museum 1880
EBC 55561 Japan

18 Shake, rattle and blow

2

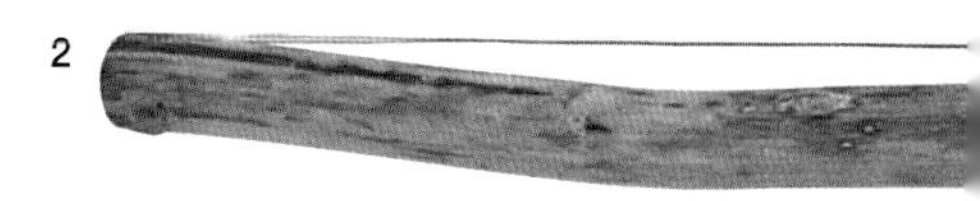

The first musical instruments may have been dried seed pods that rattled when shaken. Since then, we have found many inventive ways of making music from plants – tapping blocks of wood, blowing through hollow stems or creating resonating hollow boxes from wood or fruits.

1

Xylophone with teak sound box and bamboo notes (1)
Teak *Tectona grandis*
Donated by Paris Exhibition 1900
EBC 67606 Burma
The bamboo notes are suspended over a shaped cradle made of teak. Teak is much favoured in South East Asia for carving because it does not warp or split.

Xylophone
Strychnos sp.
Donated by A J Hipkins 1911
EBC 44836
The spherical resonators under each note of the xylophone are the fruits of a species of *Strychnos*. *Strychnos* seeds are the source of the poison, strychnine.

Drumstick
Rubber *Hevea* sp.
Donated by R Spruce 1854
EBC 44172 Brazil

Drum
Morinda angustifolia
Donated by Paris Exhibition 1900
EBC 53748 Burma
This tubular drum, made from a hollowed-out log, has skin stretched over both ends. Drums of all shapes and sizes occur all over the world, suggesting that these types of instruments are of great antiquity.

Anklet
Cayaponia kathamatophora
Collected by R Spruce 1853
EBC 54476 Brazil
The explorer Richard Spruce described how 'fruit shells are strung together and tied round the right ankle in the dabocurís (dances) of the Barré Indians, producing a loud rattling noise with every movement'.

Castanets
West Indian ebony *Brya ebenus*
EBC 60070
West Indian ebony is a small tree native to Cuba and Jamaica. Its wood is often used to make small musical instruments.

Horn covered with bark
Birch *Betula* alba
Donated by Miss Bell 1896
EBC 42476 Norway
This alpenhorn is made from a wooden log, split in half lengthwise and hollowed out to form the bore. The outside is shaped and the two halves bound together with a wrapping of birch bark.

Yagua Indian flute
Member of grass family
Donated by G T Prance 1992
EBC 71826 Peru
Grass stems of different lengths are bound together in a row. Although there are no fingerholes, the musician can produce different notes from each tube by blowing across the ends more strongly.

Reed flute
Arundo donax
Collected by P Davis 1997
EBC 73848 Mexico
This type of flute is known as a duct flute. The upper end is blocked except for a small duct that the player blows into. Covering the fingerholes in different combinations produces a variety of notes.

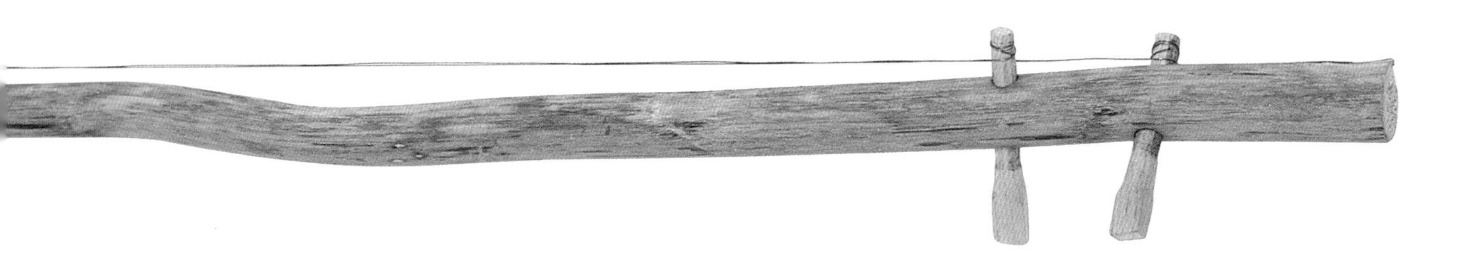

Flute
Rosewood *Dalbergia* sp.
EBC 71804 India
Many species of *Dalbergia* are valued for musical intruments such as guitars and pianos because their wood is hard and takes a fine shine. However, overharvesting is threatening the trees with extinction.

Mouth bow (2)
Agave wocomahi
Collected by P Davis 1997
EBC 73860 Mexico
This instrument is made from the flowering stem of an agave. The musician places the top of the stem in their mouth and plucks the string. The mouth acts as a resonating chamber and amplifies the sound.

Lute from storm woods
Donated by S Barber 1987
EBC 73629 Great Britain
This lute is made from timbers felled at Kew during the Great Storm in October 1987. The neck and pegboard are made of *Sophora japonica* and the back of osage orange (*Maclura pomifera*).

Nutshell whistles
Black walnut *Juglans nigra*
Donated by Dr Lombe
EBC 42685

3

Musical fruits

Gourds are the dried fruits of various members of the cucumber family, and are often used for making musical instruments. When a hollow gourd is struck it resonates, making a pleasant sound. People have used this property to create drums and rattles. Other instruments, such as mbiras and sitars, use the gourd as a sound box to amplify the noise made by vibrating parts.

Beaded rattle
Bottle gourd *Lagenaria* sp.
Donated by J Wetherell 1856
EBC 44837 Brazil
The gourd is covered with a loose net ornamented with beads. The musician holds it loosely in one hand and beats it gently while shaking it at the same time. Bottle gourds grow on sprawling, clambering vines.

Mbira (3)
Bottle gourd *Lagenaria* sp.
Donated by M S Johnson 1930
EBC 71801 Zimbabwe
The mbira consists of a series of tuned metal strips attached to a piece of wood over a gourd sound box. It produces a gentle and pleasant sound, and is often used in Africa to accompany love-songs.

Wind instrument
Bottle gourd *Lagenaria siceraria*
Bamboo
Donated by J Veitch and Sons 1894
EBC 71803 Borneo
This instrument is probably a mouth-blown organ. The musician blows into the hole at the tip of the gourd's neck and alters the notes by covering the holes in the bamboo pipes.

Scraper
Bottle gourd *Lagenaria siceraria*
Donated by Mrs Hertslet 1891
EBC 54684 Puerto Rico
A series of closely spaced grooves covers the surface of the scraper. The musician plays it by rubbing a stick up and down the gourd to make a rasping sound.

Tambura
Bottle gourd *Lagenaria* sp.
Donated by India Museum 1880
EBC 44838 India
Gourds are used as resonators in many Indian stringed instruments. They amplify the sound produced by the strings. The four strings of the tambura are played without fingering to produce a drone sound.

Maraca
Bottle gourd *Lagenaria siceraria*
Donated by R Spruce
EBC 71802 Brazil
Rattles, such as this maraca, may have been among the first musical instruments. Dried gourds are ideal rattles as their loose seeds shake around. Rattles were often used in rituals as they were believed to summon spirits.

19 Playing with plants

Many sports and games rely on plants. The willow tree gives us the cricket bat, the rubber tree the bouncing ball, and bowls are made of the dense wood of the lignum vitae tree. Other, quieter games rely on plants too: many chess sets feature carved wood pieces, as do other board games, and playing cards are made of cardboard.

2

Bowling balls
Lignum vitae *Guaiacum* sp.
EBC 40052
The lignum vitae tree, from tropical America, produces an extremely heavy, hard greenish-black timber that is prized for bowling balls. Each set of balls is cut from a single log so that they all play in the same way.

Carved chess pieces
Box *Buxus* sp.
EBC 43774
With its fine, even texture, boxwood is valued for making turned objects, such as these chess pieces, and for intricate carvings.

Carved model of lumberjacks
Crepe myrtle *Lagerstroemia indica*
Cucumber tree *Magnolia obovata*
Donated by B Halliwell
EBC 73872 Japan
This carved model was made by people in Joga village in Japan. Until recently, most of the scrap wood produced in the village was burnt for heating. Now, these toys use up the waste.

Woven rattle
Brazilian grass *Coccothrinax argentata*
EBC 35329
Although commonly known as Brazilian grass, the leaves used to make this rattle belong to a species of palm.

Toy sword
Kapok *Ceiba pentandra*
Donated by J H Holland 1897
EBC 65324 Nigeria
The lightweight wood of the kapok tree was often used to make children's toys, though the tree is more widely known for the silky mass of hairs from its fruits used as stuffing for toys and life jackets.

Wooden playing cards (1)
White spruce *Picea glauca*
Donated by M Bourgeau
EBC 27713 North America

Traditional toys

Charming toys have been hand-carved from softwood for centuries. The preliminary work was often done on a turning lathe, but such toys were always hand-finished, as in these samples made from spruce.

Toy soldiers
Spruce *Picea* sp.
Each soldier is made from a skittle-shaped wooden body formed from a single piece of wood. Arms and legs whittled from twigs were added before the soldiers were painted.

German lathe-turned toys (2)
Spruce *Picea* sp.
Germany
Toy makers in Germany produced lathe-turned rings with grooves placed so that each slice cut from the ring resembled a particular animal. Additional features were hand-carved.

1

Hoop and stick
Stick: poplar *Populus tremuloides*
Hoop: oak *Quercus* sp.
Donated by P Davis 1996
EBC 73859 Mexico
This hoop and stick are used by Tarahumara women in a hoop race called a rowéane. The coloured material covering the hoop distinguishes the two teams.

Mallet and ball used in Jeu de Mail
Handle: *Celtis australis*
Head: holm oak *Quercus ilex*
Ball: box root *Buxus* sp.
Donated by A Prior
EBC 43514 France
Jeu de Mail is probably a version of the game pall mall, played from the sixteenth century. Players hit the ball along an alley and through a hanging ring – the winner is the person with the fewest 'hits'.

Warri board
Caesalpinia bonduc
Donated by International Forestry Exhibition 1884
EBC 59426 Sierra Leone
Warri, or oware as it is sometimes called, is one of the most popular games in Africa and the Caribbean islands. Each player begins with four nickernuts per dish and aims to capture those belonging to their opponent.

Seeds and fruit – nickernuts
Caesalpinia bonduc
Donated by Royal Pharmaceutical Society 1983
EBC 58136

oy made from doum palm kernel
Doum palm *Hyphaene thebaica*
Donated by Miss Johnstone 1927
BC 35695 Egypt

Rubber ball
Rubber *Hevea* sp.
Collected by W Mukete 1995
EBC 73559 Cameroon
This lightweight ball was made by wrapping strips of rubber latex around a balloon. The first known rubber balls belonged to the Aztecs and Mayans of Central America. They were solid and bouncy.

Coutchouc ball and cup
Rubber *Hevea* sp.
Donated by Mackintosh and Co. 1853
EBC 44144
This hard ball has been made of rubber treated with sulphur. This process, known as vulcanisation, produces rubber that is proof against heat and cold and resistant to melting.

Cricket

Willow is still preferred for cricket bats because it can absorb the impact of a ball travelling at speeds of up to 150 km per hour. Other plants in cricket include ash in the stumps, cork in the ball and rubber in the protective gloves.

Cleft of willow for cricket bat
Cricket bat willow *Salix alba* var. *caerulea*
Donated by Army and Navy Stores 1895
EBC 41473

Finished cricket bat (3)
Cricket bat willow *Salix alba* var. *caerulea*
Donated by Army and Navy Stores 1895
EBC 41463

Golf

These golf clubs are made of beech wood. Other suitable timbers include hickory for the shaft and persimmon for the head. Although they are still called 'woods', golf clubs are now often made of carbon fibre and steel.

Unfinished golf club
Beech *Fagus* sp.
Donated by Army and Navy Stores 1895
EBC 41574

Golf club
Beech *Fagus* sp.
Donated by Army and Navy Stores 1895
EBC 41575

Golf ball
Gutta percha *Palaquium gutta*
Donated by F N Howes 1929
EBC 50967
Made of latex from the gutta percha tree of Southeast Asia, this type of golf ball came into use in the mid-nineteenth century. Today, golfers prefer balls made of coiled elastic wrapped in latex from chicle (*Manilkara zapota*), a tree of Central America.

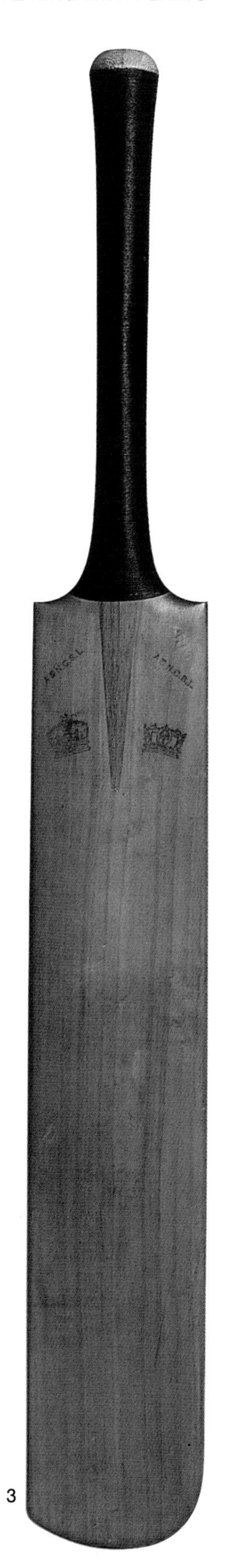
3

20 Plants for energy

Wood is still the world's most important fuel for heating and cooking. Other plant products are burnt for light: waxes for candles and oils for lamps come from a wide range of plants. In developing countries, wood supplies over 90 per cent of household energy. However, fuelwood is becoming scarcer and harder to find. Kew scientists are studying trees that could be cultivated to provide sustainable supplies in southern Africa.

1

Rush light wicks made from pith
Soft rush *Juncus effusus*
Donated by J R Evans
EBC 34582
To make rush lights, the central pith from soft rush leaves was soaked in fat or tallow. Until the middle of the nineteenth century, rush lights were used at night in many rural communities.

Candle nuts
Candle nut tree *Aleurites moluccana*
Donated by Imperial Institute 1996
EBC 73647 Malaysia
These oil-rich seeds are strung together and burnt as candles in some South East Asian countries.

West Indian castor oil
Castor oil plant *Ricinus communis*
Donated by A S Hill and Sons
EBC 44496 Virgin Islands
Castor oil has been burnt in lamps since ancient Egyptian times. During the twentieth century, it has become widely used as an engine lubricant in planes and racing cars because it can withstand high temperatures.

Tinder and fire-lighters

Fire-lighters and tinder are made of materials that catch light easily, sometimes from a single spark. Dried maize cobs or other crop wastes are ideal, and stem pith from trees such as *Aeschynomene aspera* also burns well.

Flint, steel and stem pith tinder for striking fire
Aeschynomene aspera
Donated by Paris Exhibition 1900
EBC 59917 India

Tinder made from leaf surrounding fruits
Oil palm *Elaeis guineensis*
Donated by Niger Expedition 1859
EBC 35582 Nigeria

Fire-lighters made from maize cobs
Maize *Zea mays*
Donated by W T Thistleton-Dyer
EBC 33570
William Thistleton-Dyer was Kew's third Director (1885–1905).

Rhus wax candles

The pea-sized fruits of *Rhus succedanea* contain a fatty substance that burns well and is used for candles in China and Japan. Wax cakes bearing a Japanese character were usually designed for the European market.

Wax from fruits (1)
Rhus succedanea
Donated by Japan-British Exhibition
EBC 62108 Japan

Wax candles (1)
Rhus succedanea 1885
EBC 62029 Japan

Flammable resins

Resins, such as balsams or those in pine wood, are rich in hydrocarbons, which catch light easily and burn well. They are used to start fires or, in the case of these balsam spills, for lighting church candles.

Spills filled with balsam
Balsam of Peru *Myroxylon balsamum*
Donated by T C Archer 1854
EBC 60812

The head of the Zimbabwe Forestry Commission field station at Chesa, Bulawayo, measuring *Acacia tortilis* in a trial on selected drought tolerant varieties in the Zimbabwe-Kew Fuelwood Project.

Fire-lighters dipped in pine resin
Pine *Pinus* sp.
EBC 28171 France

Wood prepared for matches and match boxes
Aspen *Populus tremula*
Donated by Paris Exhibition 1900
EBC 41347 Russia
Timber from various species of poplar (*Populus*) is the most common source of wood for matches.

Matches
Euphorbia abyssinica
Donated by P R O Bally 1951
EBC 43930 Ethiopia
Several species of *Euphorbia* produce latex containing flammable chemicals. These ensure that the stems and twigs light easily.

Vegetable tallow

Vegetable tallow is obtained from the waxy mass surrounding the seeds of *Sapium sebiferum*. It burns without smoke. Vegetable tallow candles were often coated with wax to prevent them softening in warm weather.

Vegetable tallow from seeds
Sapium sebiferum
Donated by Prices Patent Candle Co.
EBC 44605 China

Unfinished candles of vegetable tallow
Sapium sebiferum
Donated by R B Jackson 1850
EBC 44598 China

Candles made from vegetable tallow
Sapium sebiferum
Donated by R B Jackson 1850
EBC 44604 China

Charcoal

Charcoal is made by heating wood without air, so that it does not burn completely. As a fuel, it burns more cleanly than the original wood and is lighter to transport. Much commercial barbecue charcoal is made from non-renewable tropical rainforest or mangrove swamp timber. Kew has developed Bar-B-Kew charcoal, which is produced from sustainably harvested woodland at Wakehurst Place, our garden in Sussex. Wood is converted to charcoal in kilns.

Teak charcoal
Teak *Tectona grandis*
Donated by India Museum 1880
EBC 46303 India
In India and many other tropical countries, city dwellers rely on charcoal as fuel because it is easier and cheaper to transport than wood. Charcoal is also used in industry for iron smelting and brick making.

Bar-B-Kew charcoal
Ash *Fraxinus excelsior*
Great Britain
Kew is one of several organisations seeking to produce more British charcoal to meet the demand for barbecue fuel. Ash coppice at Wakehurst Place is cut on a 20–30 year cycle and the trimmings made into charcoal.

Carnauba wax

The wax palm's young leaves are coated with powdery wax – 1300 leaves produce just 1 kilogram of wax. The wax is popular for church candles because it burns well and has a pleasant scent.

Carnauba wax palm leaves
Wax palm *Copernicia cerifera*
1934
EBC 40553

Carnauba wax candles
Wax palm *Copernicia cerifera*
1882
EBC 40609 Brazil

Fuels

Thorny branch
Acacia karroo
Donated by B Hocking 1968
EBC 58847 South Africa
Acacia karroo is under investigation by scientists in Kew's Jodrell Laboratory in a project to identify drought-resistant fuelwood trees. It burns slowly and reaches high temperatures but produces very little smoke.

Tumblelog fireplace fuel
Tumbleweed *Salsola kali*
Donated by University of Arizona 1986
EBC 45966 USA
Though associated with the dry regions of south-western USA, tumbleweed was only introduced there in the nineteenth century. Scientists have devised a way of making fuel logs from its powdered stems.

Petroleum nuts (2)
Petroleum nut tree *Pittosporum resiniferum*
Donated by Royal Pharmaceutical Society 1983
EBC 60923 Philippines
These oily fruits can either be burnt directly, or be processed into a high octane fuel. During World War 2, Japanese troops powered their vehicles with fuel from the petroleum nut tree.

Para balsam
Copaiba *Copaifera multijuga*
Donated by Royal Pharmaceutical Society 1983
EBC 53401
The hydrocarbon-rich semi-liquid resin produced by the copaiba tree is very similar to diesel oil in composition. It does not need processing before being used as an engine fuel.

Fuel brickettes from shells and fibre
Oil palm *Elaeis guineensis*
Donated by F N Howes 1925
EBC 35666 Ghana
Using crop wastes as fuel not only makes the maximum use of the crop, but also avoids problems of waste disposal. Waste materials from oil palm processing are compressed into fuel brickettes.

2

21 Plants on the move

Many traditional modes of transport rely on plants. Water transport, from canoes to tall ships, has depended on trees for wooden hulls, oars and masts. Ropes and sails are woven from plant fibres and waterproofed with plant resins.

1

Rubber

All modern vehicles rely on their tyres for a smooth ride. Made of natural rubber, the tyres cushion the vehicle and grip the road surface. Rubber is made from the latex of a tree native to the South American rainforest. Today, most is harvested from plantations in Southeast Asia, and over half the world's production is used in tyre manufacture. In parts of the USA, tyre rubber is now recycled as a durable road surface. Concorde's tyres depend on natural rubber to withstand the heat and friction of touchdown.

Rubber-tapping knives and cup for collecting rubber
Donated by Stacy
EBC 44076
Using the grooved knife, rubber tappers remove a strip of bark from the trunk of a rubber tree. Rubber latex exudes from the slit and is collected in a small cup, then put into metal containers for transport.

Gourd used for collecting fluid rubber
Donated by A Johnston 1885
EBC 44208 Brazil

Latex coagulated with acetic acid
Rubber *Hevea brasiliensis*
Donated by Royal Pharmaceutical Society 1983
As it exudes from the tree, rubber latex is a white liquid. Acetic acid is added to the latex to coagulate the rubber particles into a solid mass.

Smoked sheet rubber
Rubber *Hevea brasiliensis*
EBC 44162 Sri Lanka
Once the coagulated rubber has been rolled into sheets it is placed in a smoke house. The high temperatures dry the sheets out and the smoke acts as a preservative, preventing mould growth.

Graded samples of rubber
Rubber *Hevea brasiliensis*
Donated by Malayan Rubber Industry 1996
EBC 73626 Malaysia
The Malayan Rubber Industry developed a way of reducing solidified rubber to crumbs by adding castor oil, another plant product, and passing it through rollers. The crumbs are then compressed into blocks.

Para rubber (1)
Rubber *Hevea* sp.
Donated by C Mackintosh and Co. 1853
EBC 44242 Brazil
Rubber tappers harvesting from wild trees in Amazonia still prepare their crude rubber for sale by pouring the liquid latex onto a rotating stick over an open fire. The smoke and heat solidifies the rubber.

High performance car tyre
Rubber *Hevea brasiliensis*
Donated by Silverstone Tyre and Rubber Co. SDN BHD 1997
Natural rubber is better than synthetic alternatives for tyres because it is stronger, has greater resistance to impact and is relatively insensitive to temperature changes.

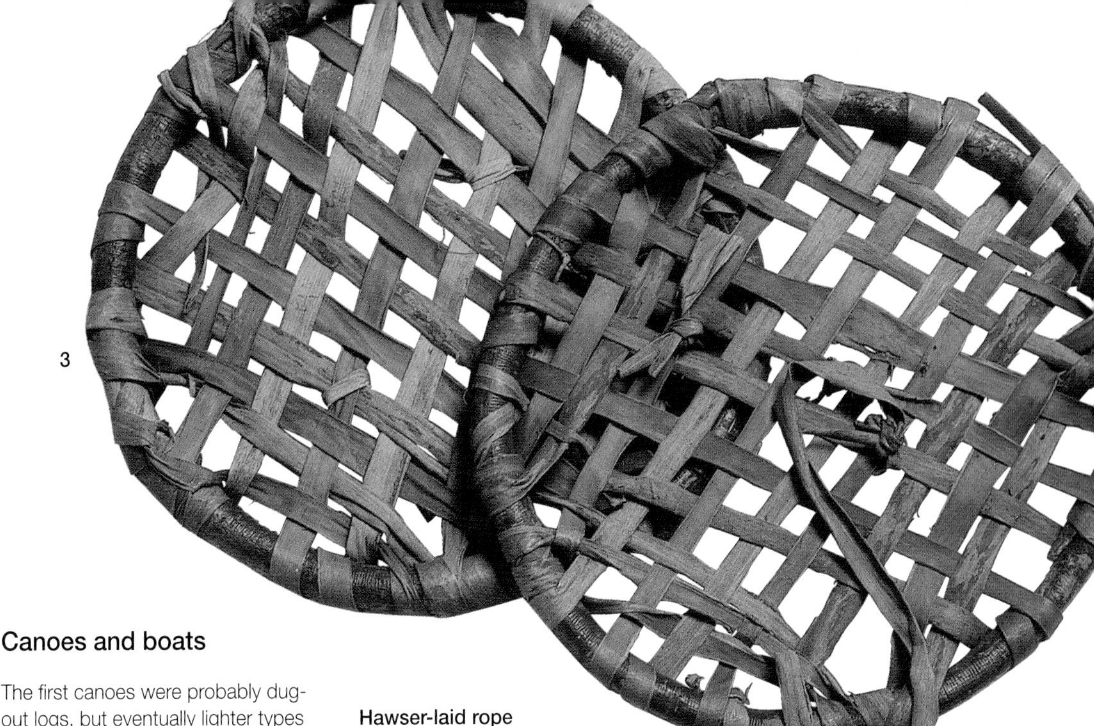

3

Germinating rubber seeds
Rubber *Hevea brasiliensis*
Collected by H A Wickham 1876
EBC 44029
In 1876, Henry Wickham sent 70,000 rubber seeds from Brazil to Kew. Less than 2,500 germinated but those seedlings were sent to Sri Lanka and Malaysia, forming the basis of the Far Eastern rubber industry.

Paddles

Many different woods are used to make paddles for canoes and other boats. In Guyana, the wood of one tree, *Aspidosperma excelsum*, was prized so highly for this purpose that it is known as the paddlewood tree.

Paddle
Paddlewood tree *Aspidosperma excelsum*
Donated by Exhibition of Works of Industry, British Guiana
EBC 67690 Guyana

Paddle
Oregon maple *Acer macrophyllum*
Donated by C F Newcombe 1904
EBC 61597 North America

Wooden paddle
Ash *Fraxinus floribunda*
Donated by India Office 1893
EBC 67672 India

Canoes and boats

The first canoes were probably dug-out logs, but eventually lighter types were constructed. Birch bark canoes had a single piece of bark stretched over a cedar frame and stitched with white spruce roots.

Model of canoe
Paper birch *Betula papyrifera*
Donated by Dr McGillivray 1848
EBC 55000 Canada

Model of canoe (2)
Poplar *Populus* sp.
Donated by Colonial and India Exhibition 1886
EBC 41358 Canada

Model of fishing boat
Teak *Tectona grandis*
Donated by J Henshall 1852
EBC 73324 Java
This Javanese fishing boat had a hull constructed of teak, masts and yards of bamboo, and sails and ropes made from the fibre of the gebang palm (*Corypha gebang*).

Wooden scoop for bailing canoes
Calophyllum sp.
Donated by E H Man 1901
EBC 66544 India

Hawser-laid rope
Sisal *Agave sisalana*
Donated by Death's Fibre Machine Co. 1890
EBC 29928 Cuba
The long narrow leaves of the succulent Central American plant, *Agave sisalana*, are an important source of natural rope-making fibres. Sisal ropes are strong enough to be used on ship's rigging and anchor cables.

Snow shoes

People wear snow shoes to distribute their weight over a larger area so that they are less likely to sink into soft snow. The woven surface of the shoes improves the grip on slippery snow.

Snow shoes (3)
Hazel *Corylus avellana*
Donated by Hungary Exhibition 1908
EBC 42338 Hungary

Snow shoes
Rock maple *Acer glabrum*
Donated by C F Newcombe 1904
EBC 69756 USA

Model of Indian country cart
Teak *Tectona grandis*
EBC 37962 India

Bridle of cotton cord and tape
Cotton *Gossypium* sp.
Donated by India Museum 1880
EBC 65615 India

Yak saddle
Rhododendron hodgsonii
Donated by J D Hooker
EBC 67662 Himalaya
Sir Joseph Hooker, was an intrepid plant collector who introduced many rhododendrons into cultivation. This yak saddle, which he used in the Himalayas, is made of rhododendron wood.

2

22 To dye for

Plants provide a rainbow of coloured dyes. Some plant extracts create different colours depending on the other chemicals used in the dyeing process. We can use these colours in many ways to decorate our clothes and ourselves – body-painting and hair-dyeing are common across many cultures.

1

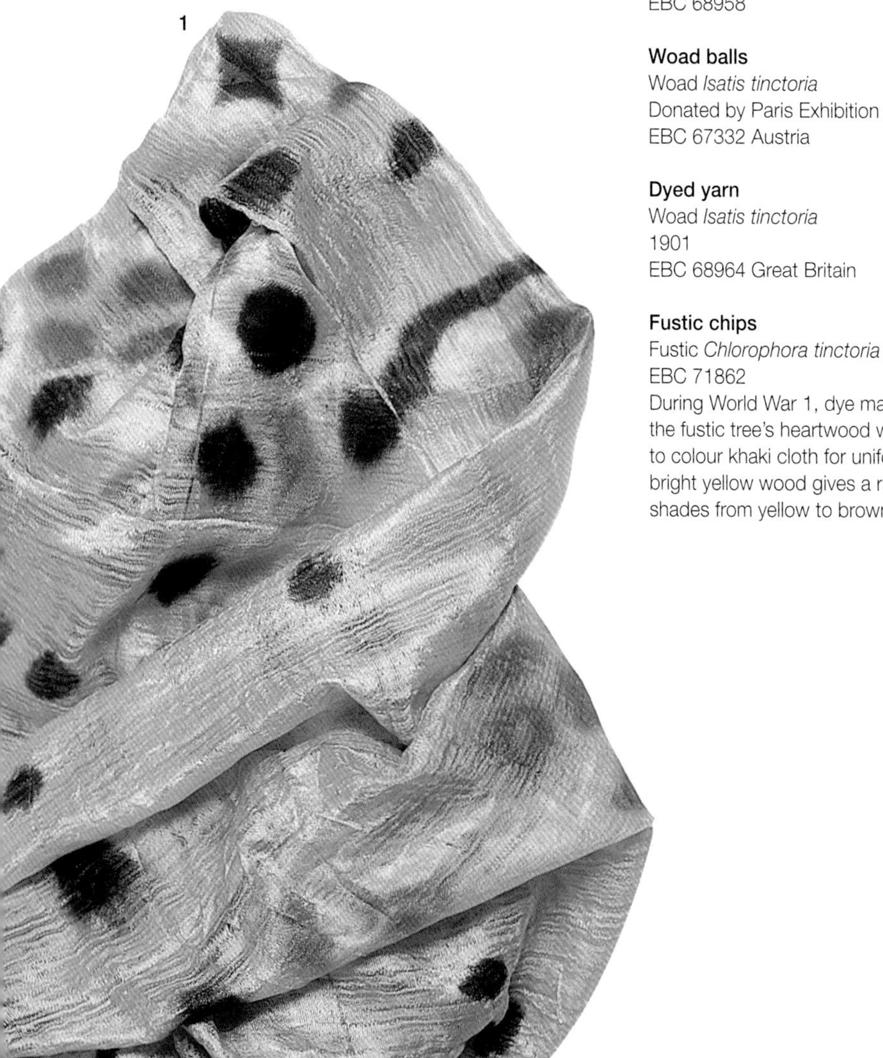

From Celtic body paint to uniform dye

In 54 BC Julius Caesar described how British warriors stained themselves blue using a dye made from woad leaves. The same dye was used until 1932 to colour police uniforms

Woad leaves
Woad *Isatis tinctoria*
EBC 68958

Woad balls
Woad *Isatis tinctoria*
Donated by Paris Exhibition 1885
EBC 67332 Austria

Dyed yarn
Woad *Isatis tinctoria*
1901
EBC 68964 Great Britain

Fustic chips
Fustic *Chlorophora tinctoria*
EBC 71862
During World War 1, dye made from the fustic tree's heartwood was used to colour khaki cloth for uniforms. The bright yellow wood gives a range of shades from yellow to brown.

Silk dyed yellow (1)
Lodh *Symplocos racemosa*
Donated by J D Hooker
EBC 50446 India
Both the bark and leaves of the lodh tree are used in India to produce yellow dyes.

Woven sampler of natural dyes
EBC 73269
The yellow and brown dyes that colour this sampler have all been produced from various species that grow in Britain including meadowsweet (*Filipendula ulmaria*) and weld (*Reseda luteola*).

Safflower
Safflower *Carthamus tinctorius*
EBC 68965 India
A yellow dye from safflower petals is used to dye the robes of Buddhist monks. They can also yield a red dye which once coloured the cotton tapes binding legal documents – hence the term red tape.

Saffron
Saffron *Crocus sativus*
Donated by Royal Pharmaceutical Society 1983
EBC 37050
Saffron comes from the hand-picked bright orange stigmas (part of the female reproductive organs) of the saffron crocus flower. They produce a yellow dye and are also used to flavour food.

Wool dyed with onion-skins
Onion *Allium* sp.
EBC 40594
The papery brown scales wrapped round onion bulbs produce a range of yellow and orange colours.

2

4 & 3

Making indigo

True blue

Indigo puts the blue into jeans. The green shoots of the indigo plant produce the vibrant dye used to colour denim. Indigo is native to India, but is now farmed in many other hot countries. For a long time, it was a valuable trading item between India and the rest of the world, and was known as the 'king of the dyestuffs'. In the first stage of indigo production, the crushed leaves are steeped in cold water.

Block indigo
Indigo *Indigofera tinctoria*
Donated by India Museum 1880
EBC 55450 India
In water the indigo plant releases a colourless substance that turns blue when beaten and exposed to the air. This blue pigment settles out as a sediment which is dried into blocks.

Indigo tablets (2)
Indigo *Indigofera tinctoria* 1852
EBC 73245

Native dress of indigo-dyed cotton
Indigo *Indigofera* sp.
Donated by F W Meres 1898
EBC 65617 India
Several species of *Indigofera* are cultivated as dye-plants throughout the tropics. However, most indigo dyes used today are artificial.

Jeans
Indigo *Indigofera* sp.
Donated by P Bowers 1997
Natural indigo is still used to dye denim for certain brands of jeans. It produces a long-lasting hard-wearing colour. However, much blue fabric is now dyed with synthetic indigo, first developed in 1887.

Indigo balls
Indigo *Indigofera tinctoria*
Donated by Dr Baikie 1859
EBC 60541 Nigeria
Indigo leaves are picked when they are fully developed and torn or chopped to release the juice. The pulp is made into balls and dried for storage. Broken up in water, the balls release the dye.

Body paint and cheese colour

Annatto comes from *Bixa orellana* seeds. Throughout Amazonia, people use it as a body paint and hair dye. Elsewhere, it's a food colouring (E160b) in cheese, margarine and other dairy products.

Annatto fruits and seeds
Annatto *Bixa orellana*
Collected by N Rumball 1995
EBC 73532 Cameroon

Ball of annatto
Annatto *Bixa orellana*
Donated by International Exhibition 1862
EBC 67045 Guyana

Henna leaves
Henna *Lawsonia inermis*
Donated by India Museum 1880
EBC 54968 India
Henna is a red-brown skin stain and hair dye popular throughout India, North Africa and the Middle East. The Berbers believe it represents fire and blood, linking mankind with nature.

Holi powder
Palas tree *Butea monosperma*
Donated by India Museum 1880
EBC 54883 India
During the 'Holi' religious festival, Hindus celebrate with dancing and processions and people fling Holi powder. This powder contains a pigment made from the palas tree's yellow flowers.

Black wood, blue dye

Once known as black wood, the heartwood from this tropical tree was one of the most important dye sources found in the Americas by European explorers. It produces a variety of dyes from black to blue.

Logwood
Haematoxylon campechianum
Donated by A S Hill and Son
EBC 71864

Logwood extract
Logwood *Haematoxylon campechianum*
EBC 59720

Madder

Making turkey-red dye from madder roots was long and complicated. Twenty separate stages involved blood, oil and rancid fat! Madder was used for hunting coats known as 'hunting pinks'.

Madder powder (3)
Madder *Rubia tinctoria*
Donated by Great Exhibition 1851
EBC 68959 Spain

Wool coloured with madder (4)
Madder *Rubia tinctoria*
Donated by Quantock Weavers
EBC 53899

Brazilwood

Brazilwood from various species of *Caesalpinia* yields a red-brown dye. European explorers recognised brazilwood trees in South America and named the area 'Brazil' after the tree.

Brazilwood chips
Brazilwood *Caesalpinia echinata*
Donated by Royal Pharmaceutical Society 1983
EBC 61702 India

Ground brazilwood
Brazilwood *Caesalpinia echinata*
Donated by W Gourlie
EBC 59375

23 Where to go in the gardens

You'll encounter a wide range of useful plants when you visit Kew's gardens and glasshouses. These diverse objects represent just some of the source plants on display – look out for other plants featured in the **Plants+People** exhibition.

1

Princess of Wales Conservatory

Frankincense resin
Frankincense *Boswellia sacra*
Donated by M van Slageren 1997
EBC 73880

Temperate House

Desert Whale – jojoba oil (1)
Jojoba *Simmondsia chinensis*
Donated by University of Arizona 1985
EBC 71974

Jojoba seeds (2)
Jojoba *Simmondsia chinensis*
Donated by University of Arizona 1985
EBC 43805

Waterlily House

Lotus root starch
Lotus *Nelumbo nucifera*
1982
EBC 41215

White lotus nuts
Lotus *Nelumbo nucifera*
1984
EBC 41198 China

Order Beds

Hyoscine tablets
Henbane *Hyoscyamus* sp.
EBC 55201

Grass Garden

Brush
Sorghum bicolor
Donated by H Battock
EBC 32527 Egypt

Palm House

Ebony dish
Ebony *Diospyros ebenum*
Donated by Paris Exhibition
EBC 73880

2